Solar Projects

Working Solar Devices To Cut Out and Assemble

A. Joseph Garrison

RUNNING PRESS
PHILADELPHIA, PENNSYLVANIA

9 8 7 6 5 4 3 2 1
Digit on the right indicates the number of this printing.

Canadian representatives: John Wiley & Sons Canada, Ltd.
22 Worcester Road, Rexdale, Ontario M9W 1L1.
International representatives: Kaiman & Polon, Inc.
2175 Lemoine Avenue, Fort Lee, New Jersey 07024

Library of Congress Cataloging in Publication Data:

Garrison, A. Joseph.
Solar projects.

Bibliography: p.
1. Solar energy. I. Title.
TJ810.G36 621.47 81-5130
ISBN 0-89471-129-6 (lib. bdg.) AACR2
ISBN 0-89471-130-X (pbk.)

Edited by John Prenis
Cover art direction by James Wizard Wilson
Cover photography by Carl Waltzer

Typography: Gill by Graphic Dimensions, Inc., Pennsauken, New Jersey
Printed by Port City Press, Baltimore, Maryland

This book may be ordered directly from the publisher.
Please include 75 cents postage.

Try your bookstore first.

Running Press
125 South Twenty-Second Street
Philadelphia, Pennsylvania 19103

Contents

Acknowledgements

National Solar Heating and Cooling Information Center
P.O. Box 1607
Rockville, MD 20850
For Mean Solar Radiation Charts and Map.

Dr. Al Hughes and Dr. Kermit Horn
For testing projects in numerous workshops throughout Oregon.

Students at Churchill High School
Who inspired many of the ideas in this book.

Lane Education Service District
For graphic assistance in some projects during teacher inservice workshops.

Introduction

The consumption of energy throughout the world is depleting our natural resources at an alarming rate; we are eating away at the very planet we live on. At the present time between 95 and 98 percent of our energy comes from fossil fuels: coal, oil and natural gas. No one is sure just how long these supplies will last. We do know, however, that the source is limited; it will run out sooner or later.

The time to plan for the future is now. Solar energy is but one of the many possibilities that should be studied as an alternative to our present needs. It is a dependable energy source; the sun is expected to last at least four billion more years. It strikes the earth in huge quantities each day. Unlike other energy resources, the worldwide use of solar radiation would have a minimal impact on the environment and can be renewed each day with a fresh supply.

The *Solar Projects Handbook* is designed to let the reader explore some practical solar energy uses through a hands-on approach. It provides easy-to-construct, yet workable activities for families, students and teachers that are fun to assemble and instructive. These projects are intended for both experimental and educational use; the reader should feel free to revise, modify or otherwise change their design to improve performance whenever possible. The projects presented will perform differently in various locales throughout the country because of climatic conditions. Wind, clouds, latitude and time of year are a few of the variables which may affect performance. The most important goal is the learning experience involved; if this occurs, this book will have achieved everything it was intended to.

How To Use This Book

First, look it over. The book was designed in a way that would maximize its effectiveness. The first sixty-four pages of descriptions and assembly instructions for the projects are identified by number. The rest of the book contains the plans for the projects. The plans are numbered, too, and each plan's number should be matched to its corresponding number within the first section.

Next, tear it up. We mean it. Once you pay for a book, you can do anything you want with it, right? So, grasp the project section firmly and tear it out. The projects in this book are meant to be cut out and put together. They lose all their value if you leaf through the book a couple of times and lay it aside intact.

We hope what you learn in building these projects will encourage you to improve upon them, to make bigger and better versions, and above all, to have fun.

To Jodie, Todd and Carol

PART ONE
Construction Techniques

Hints for Making Projects

Some general suggestions, instructions and tips that could be helpful when constructing most projects are as follows:

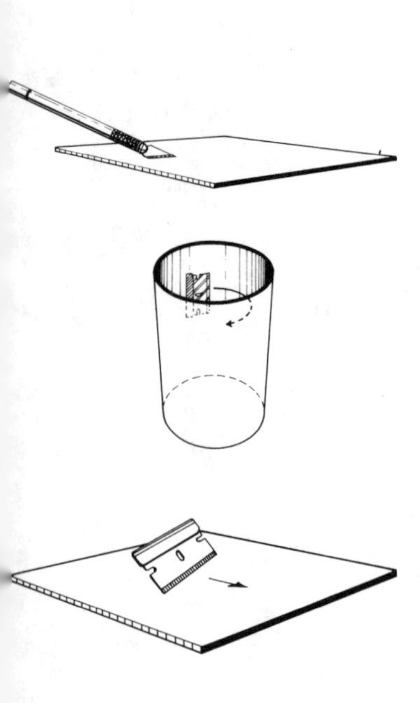

1. BE SAFETY CONSCIOUS!
   If you are not careful, it is easy to make mistakes when using sharp tools. It is also safer to use sharp tools rather than dull ones. X-acto knives, pocket knives and single-edged razor blades can all be re-sharpened by the following procedure:

   A. Use silicon carbide paper 400 grit (wet). If no silicon carbide paper is available, use Ajax cleansing powder mixed with a little water; use this on an old piece of leather to hone the edge.

   B. Hone a single or double-edged razor blade on the inside of a drinking glass. The author once managed over 70 shaves on an old Gillette Blue Blade with this technique.

   C. Use a scrap piece of leather (old belt or shoe tongue). A small amount of red rouge (jewelry polish) rubbed on the leather will allow the sharp edge to glide more easily along the pattern guidelines.

2. FOLLOW INSTRUCTIONS CAREFULLY.
   Follow the order in which instructions are presented. Take your time; be accurate with all cuts. A slight deviation from the given pattern on some projects may result in a reduction in the overall effectiveness of the project.

3. AVOID THE SUN'S GLARE.
   Several projects will use highly reflective surfaces to focus the sun's rays toward another object or surface; extreme care must be taken to avoid staring at the sun's rays directly. Severe eye damage could occur without adequate eye protection. Sunglasses are a good idea.

4. USE CARE WHEN MAKING CUTS.
   To avoid cutting the worktable, place a heavy piece of cardboard or an old magazine under the pattern when cutting with a razor blade or x-acto knife. Also, never cut toward your fingers if at all possible.

5. FOLLOW LINE SYMBOLS CLOSELY.
   A. Lines to be cut out are coded with a solid line (______________).

   B. Lines that need to be folded down are called ridge folds; they are scored (cut) only halfway through on the top and are coded with a dashed line (‒ ‒ ‒ ‒ ‒ ‒ ‒).

   C. Lines that are folded up are called valley folds. They are scored halfway through from the back side and are coded with a dotted line (. . . . .).

REMEMBER: All folded lines should be cut only halfway through the paper; be extremely careful not to cut completely through.

6. ALLOW FOR CARDBOARD THICKNESS.
Since the projects in this book may be adapted to a wide variety of conditions and materials, including enlarging some projects as the reader desires, the thickness of the cardboard used may cause some minor discrepancies to occur when assembling the finished model. Use your own judgement; make corrections as problems arise.

Tools and Materials Needed

The following tools have proven to be adequate for most of the cutout projects contained in this book, although not every tool is needed for each project. Substitutions, of course, can be made whenever necessary. The tools are:

1. Sharp, pointed scissors
2. Tin snips or metal-cutting shears (useful for cutting thick cardboard when the reader wants to enlarge a project)
3. X-acto knife
4. Single-edged razor blade
5. Paper punch
6. Stapler
7. Metal-edged ruler
8. Magnetic hand compass
9. Ponce wheel
10. Cutting board

The following materials will enable most cutout projects to be constructed without difficulty:

1. Spray adhesive
2. White glue
3. Rubber cement
4. Masking tape
5. Reflective (aluminized) mylar
6. Aluminum foil
7. String or heavy thread
8. Pushpins
9. Paper fasteners
10. Transparent tape
11. Railroad board, illustration board or other lightweight cardboard, in case projects need reinforcing or enlarging.
12. Straight pins
13. Transparent plastic (mylar) or clear self-basting oven bag
14. Metal clothes hanger
15. Flat black spray paint

All materials can be purchased locally. If necessary, use your own judgement when making substitutions; many new products on the market may be better than those listed here.

Line Symbols

The different line symbols occurring throughout this book are used to simplify construction of the projects; care must be taken to interpret drawings carefully. The lines used are:

1. ________________________________

   Outline of Project

2. ▬ ▬ ▬ ▬ ▬ ▬ ▬ ▬ ▬ ▬ ▬ ▬ ▬ ▬

   Ridge Fold Line (fold down)
   Score on the line.

3. . . . .

   Valley Fold Line (fold up)
   Score on the underside of the line.

Scoring

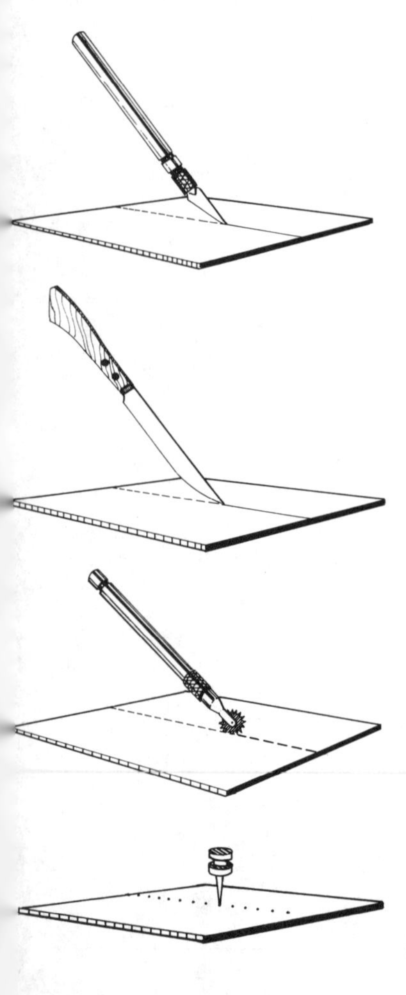

Scoring refers to the process of cutting only partially through paper or cardboard to make sharper bends and folds. Four different scoring techniques can be used for cutout projects in this book. They are:

1. Lines can be scored with a sharp razor blade or x-acto knife; care must be taken to cut no more than halfway through the paper or cardboard.
2. Lines can be scored with a dull, smooth round-tipped table knife (without teeth). This process compresses the paper rather than cutting through it.
3. Lines can be scored with a ponce wheel similar to those used in dressmaking. This is a good way to score valley folds, since the lines can be scored directly on the line rather than from the reverse side where visibility is poor.
4. Lines can be scored by pricking a sharp pin through the paper; this procedure works well for small valley folds. When using this technique, it is best to use a straight pin with a finger cap on it or a pushpin.

To obtain the best results, the following rules should be adhered to when scoring lines:

1. If a sharp instrument is used, place a heavy piece of cardboard or magazine under the paper to prevent damaging the cutting board or table.
2. For neater, more precise folds, all lines should be scored before attempting to fold them.

3. Use a metal-edged ruler as a guide when scoring straight lines.

4. When lines are to be scored on the reverse side of the printed page, either use a ponce wheel or prick the folding edge with a pushpin to reference the line on the underside.

Enlarging a Drawing

Whenever drawings in this book are to be enlarged to a greater size, it can be done quite simply by re-drawing the original onto a grid of the size desired. For example, if the following shape were to be enlarged:

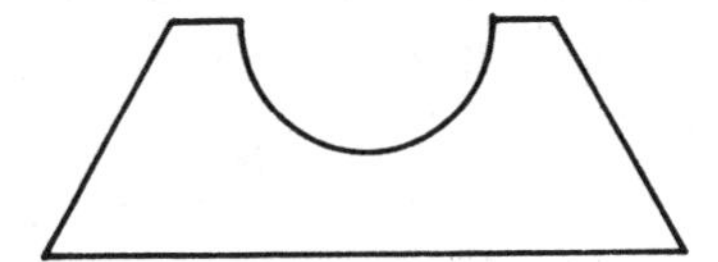

First draw a grid pattern composed of ½" squares onto the drawing, then draw a larger grid pattern on another sheet at the increased scale desired. Next, proportionately relocate each line from the original drawing. The size of the larger grid, of course, depends upon the required size of the enlargement.

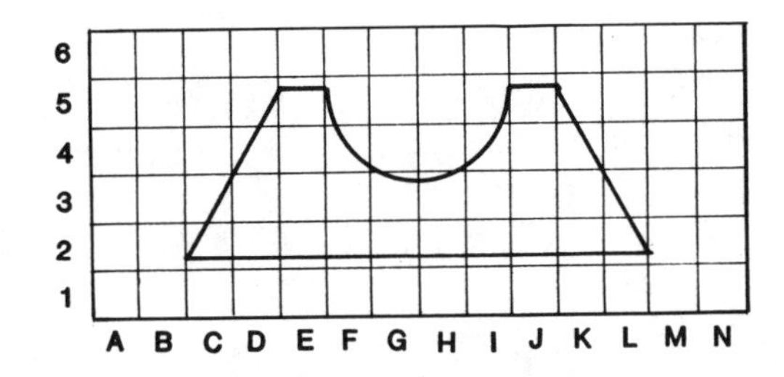

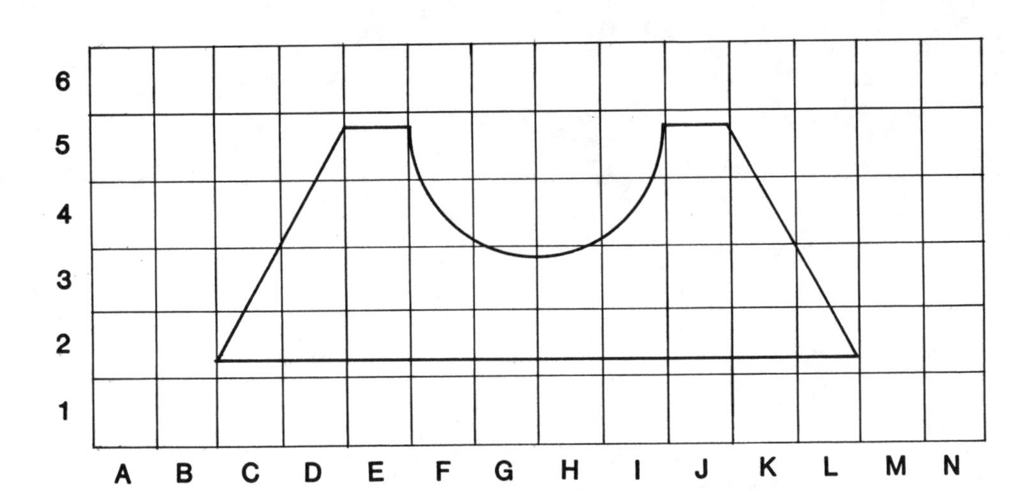

Gluing

Spray adhesives such as Scotch's "Photomount" and 3M's "77" can be extremely handy when mounting the aluminized mylar and the aluminum foil onto a rigid backing. It provides an even coating of the adhesive, permitting the glued sheets to lie flat.

Before using it, spread some newspapers on the floor or table to prevent overspray from coating the work area. Next, spray a thin, uniform coat onto the back of the paper being mounted. When the adhesive has dried, align the paper to the cardboard and press the two pieces firmly together. Next, rub the surface gently from the middle outward with a smooth cloth to eliminate air bubbles and wrinkles. If the surface is to be curved or bent, as when constructing the parabolic curve of the concentrating reflector, attach the paper after the curve has been formed. This reduces the likelihood of wrinkles occurring.

Rubber cement works in a similar manner to the spray adhesives, since it is a form of contact cement. The major problem with using it is to get a smooth, even coating; this can be done by using a cardboard straightedge as an applicator to spread it evenly over each surface. Allow both surfaces to dry completely before attaching them together.

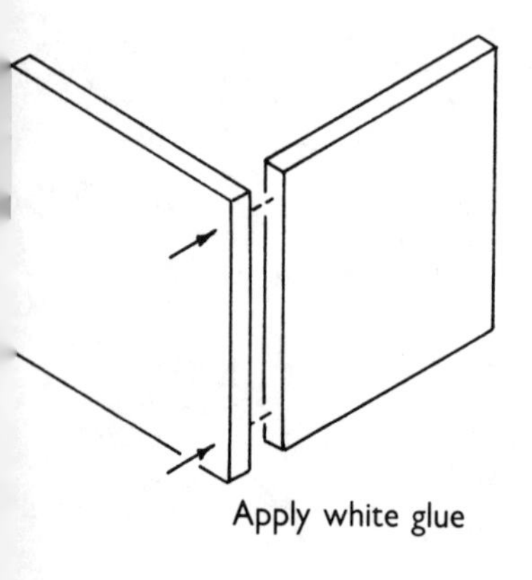

Apply white glue

White glue works well when joining edges together; the glue dries quickly when absorbed by the rigid paper, requiring only a short holding time. After applying the glue to both edges, wipe most of it off, then press the two pieces together immediately. After a minute or two the bonded edge should be strong enough to set aside for drying.

PART TWO
The Sun

The Sun: It All Starts Here

The sun is a powerful body of energy: it produced all the fossil fuels in present use, continues to produce the winds and tides as well as sustaining life on this planet, and controls the entire solar system. According to Melvin Caskin, a Nobel Prize winner and expert on photosynthesis, "The quantity of solar energy striking the earth's surface in just ten days is the equivalent of this planet's total reserves of fossil fuels." The sun, sometimes called the most ideal energy source of all, is a medium-sized star that is about 100 times the diameter of the earth, or 864,000 miles across. More than one million earths would fit into the sun. According to one solar authority, one ball of coal the size of the sun would burn up completely in 3,000 years, yet the sun has already burned for three billion years—and is expected to last yet another four billion years. The temperature at its surface is 10,000° Fahrenheit (5540° Celsius) and at its center, approximately 20,000,000° Celsius.

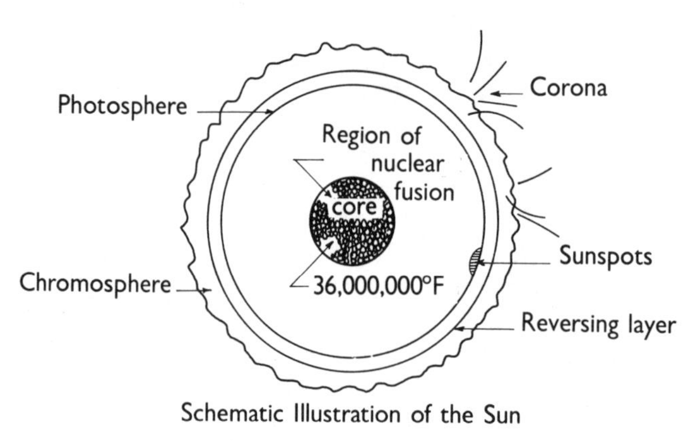

Schematic Illustration of the Sun

Ancient civilizations worshipped the sun as a god. To the Egyptians, Ra, the sun-god or god of creation, represented the sustainer of life. To them, the sun provided a sense of stability and security as it travelled across the sky each day. Ancient Egyptians sometimes referred to the sun as a giant scarab-beetle. Since they had observed the beetle pushing a round ball of dung on the ground in front of him so that he could deposit it in some crack or crevice, they reasoned that a larger beetle pushed the round sun across the sky in a similar manner.

Modern civilizations now know a great deal more about the sun and its potential uses than did our ancestors; this, in part, is due to the many contributions made by people such as Copernicus, Galileo and Newton as well as countless others. The sun's rays strike the earth in the form of solar radiation, or insolation, as it is sometimes referred to. This electromagnetic radiation must travel a distance of 92,900,000 miles before it reaches the earth. It arrives approximately 8½ minutes later in various forms of radiant energy of different wavelengths: gamma rays, X-rays, ultraviolet light, visible light, infrared light, short and long radio waves. The most common unit of measurement for solar radiation is given in langleys per minute:

1 langley = 1 calorie of radiant energy per square centimeter
1 langley per minute = 221 Btu per square foot per hour

One Btu (British thermal unit) is the amount of energy needed to raise the temperature of one pound of water one degree Fahrenheit.

Even though a constant amount of solar energy strikes the earth above the atmosphere, only about one-half of it actually reaches the earth's surface; the other half is either reflected back into space from the atmosphere or is absorbed or scattered by molecules within the atmosphere itself. Not all of the radiation reaching the earth's surface is directly from the sun; some of it is scattered, or diffuse, coming in from all directions. In cloudy weather most of it is scattered, greatly reducing the amount of usable radiation available for energy conversion.

Varying amounts of solar radiation are received at the earth's surface in different locales throughout the world—and even at the same location for different times of the year. For example, the average amount of insolation reaching Corvallis, Oregon, on December 12 is 39 langleys and on July 12, 655 langleys. The greatest amount received on earth is along the equator, which is due to the sun being almost directly overhead. As one goes north or south the amount of annual sunlight becomes less, with a corresponding decrease in the amount of solar radiation received. This amount is dependent upon the time of day and year, latitude, climatic and atmospheric conditions, physical obstructions, and altitude.

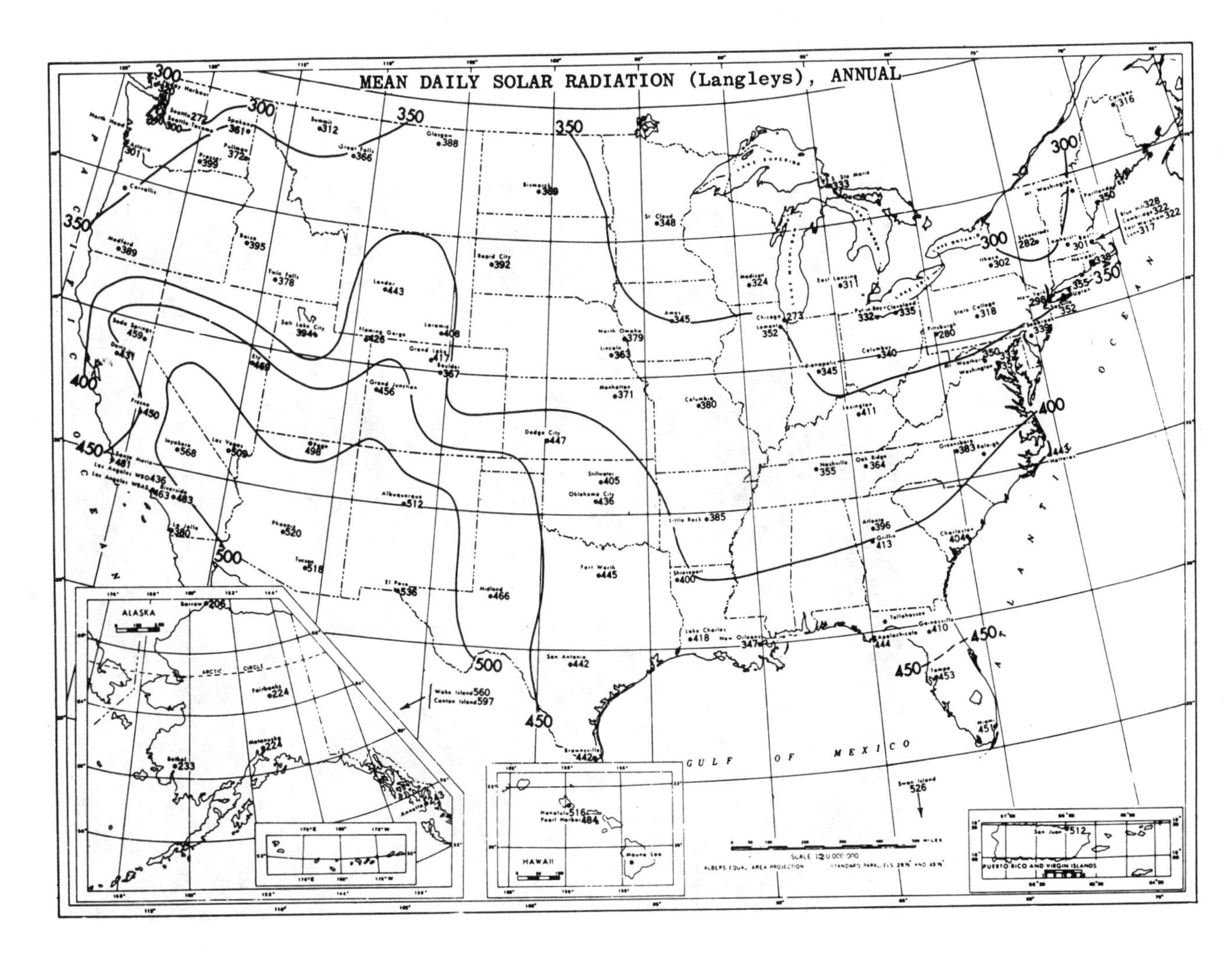

PROJECT
1

Sun Angle Viewer

The Sun Angle Viewer enables the reader to determine both the horizontal and vertical angles of the sun so that all solar projects can be oriented correctly. In addition, it can be used for other purposes in which the sun angle is needed; this includes both educational and scientific applications. For instance, if the vertical and horizontal sun angles for 11 a.m. on May 15 are needed, observe the sun angles directly at that time.

The Sun Angle Viewer works by casting a shadow that shows by its length, the vertical angle of the sun and by its direction, the horizontal sun angle.

The Sun Angle Viewer and the sundials to be described later need to be oriented with respect to true north. True north must never be confused with magnetic north. A hand compass has a magnetic needle that swings freely on a pin point; when it comes to rest, the needle will point toward magnetic, not true north. The magnetic variation of an area will be dependent on both the local magnetic attraction and its direction from the magnetic pole; U.S.G.S. maps include this information on them. When the magnetic variation is known it can then be offset on the hand compass to find true north. Many good compasses have a mechanism to adjust the dial plate so that the sighting device will point toward true north. Other compasses must be offset mathematically by adding or subtracting the local magnetic variation from the bearing shown by the needle. As a precaution, consult the instructions with your compass before adjusting it.

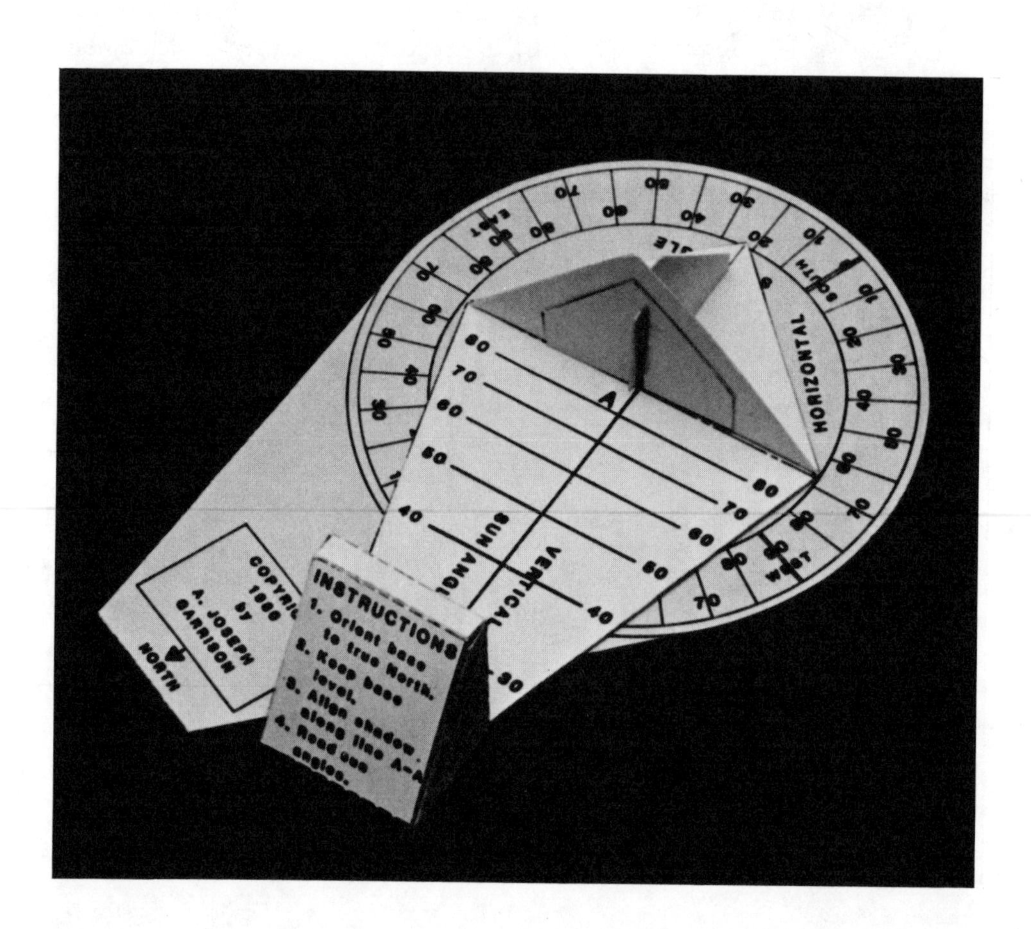
HORIZONTAL
VERTICAL
INSTRUCTIONS
1. Orient base to true North.
2. Keep base level.
3. Align shadow along line A-A
4. Read sun angles.
A. JOSEPH GARRISON
NORTH

Assembly Instructions

1. Cut out all pieces on solid (————————) lines.
2. Punch holes (+) with a pushpin (pieces A and B).
3. Score ridge folds (▬▬▬▬▬▬▬) with an x-acto knife (piece C).
4. Score the valley fold on piece D from the back side.
5. Cut out the notch (▬▬▬▬▬▬▬) on piece D and the slot on piece E with a sharp x-acto knife.

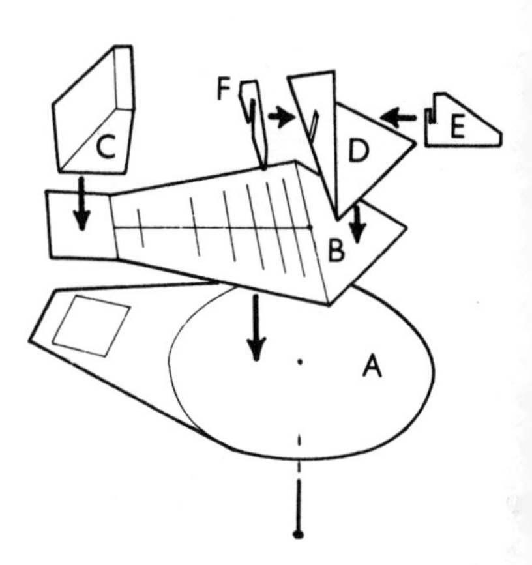

6. Fold, assemble and glue piece C.
7. Glue piece C to piece B on the rectangle marked "glue piece C here"; be sure to align axis A–A with the corresponding axis line on piece C.
8. Fold piece D at a 90° angle then glue it to the corresponding triangle on piece B.
9. Insert the vertical lock, piece E, into the slot on piece D, then glue them securely together.
10. Insert the vertical lock key, piece F, into the slot on piece E, then glue together; this locks piece E in place and positions the vertical angle pointer in an upright position.
11. Insert a straight pin into the base (piece A) from the bottom side then attach the upper assembly to it by inserting the pin into the pinhole on piece B. Keep the pin upright and apply glue to hold it in place. This completes the assembly of the Sun Angle Viewer.

Operating Instructions

1. Align the base in a true north-south direction; be sure to offset the magnetic variation on your hand compass.
2. Keep the base level or the vertical angle will not be accurate.
3. Rotate the pointer until the tip of the sun's shadow is aligned along axis A–A.
4. Read the vertical angle at the tip of the shadow and the horizontal angle from the pointer facing the sun.
   Remember: The sun is moving constantly (the sun is actually stationary, but the earth is rotating on its axis as it circles the sun). The two sun angles are continually changing. Therefore, the pointer needs to be re-oriented as the tip of the shadow moves from line A–A.

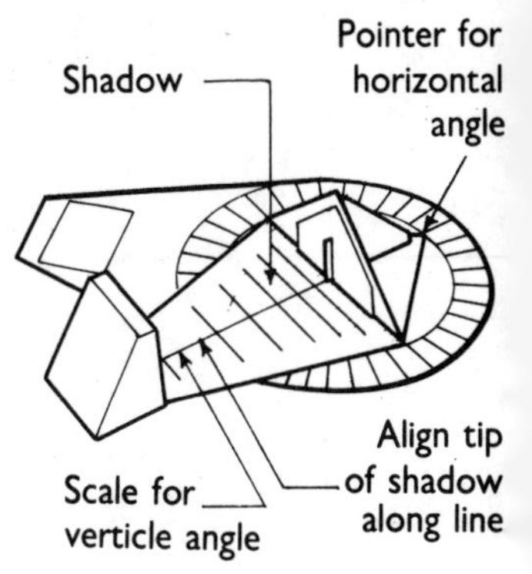

Experiments

The Sun Angle Viewer can be used to find the time of local apparent noon, when the sun is highest in the sky. This will probably *not* be the same as noon by the clock. For this experiment the Sun Angle Viewer need be pointed only roughly north. Starting about ten minutes before (clock) noon, record the sun angles each minute. When the vertical angle is greatest (and the shadow shortest) local apparent noon has occurred. At the time of local noon the shadow is pointing in a true north–south line. You might want to record the direction of true north obtained this way for future reference, and compare it with your compass.

Record the vertical angle of the noon sun each week and you will see a slow but steady shift as the seasons pass. (The measurement can be done at some other time of day if that is more convenient, as long as it is the *same* time of day.) The ancients kept track of the seasons this way.

Sundials

Sundials have been used by mankind for centuries to determine the time of day; only in recent times have we turned to more precise instruments for general use. The history of the sundial dates back to approximately 2,000 BC when it is believed to have been used in ancient Babylon. Later, in 300 BC, a Chaldean astronomer named Berossus proposed a sundial in the form of a hollow hemisphere with a gnomon and an arc that was divided into twelve equal parts. This divided the day into the Greek equivalent of our present time measurement. During Julius Caesar's time many types of sundials existed; Vitruvius was disheartened at the time because he was unable to invent a new one.

Sundials measure the position of the sun. A sundial normally works by having one part (called the gnomon or style) cast a shadow on another part (the dial plate). The dial plate is divided into equal parts so the position of the shadow can be used as a measure of time.

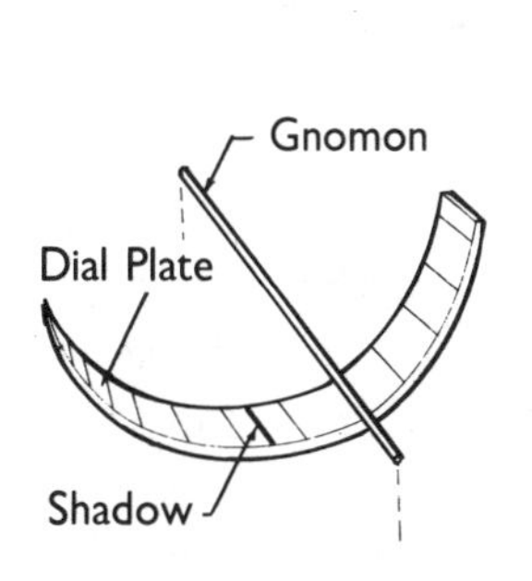

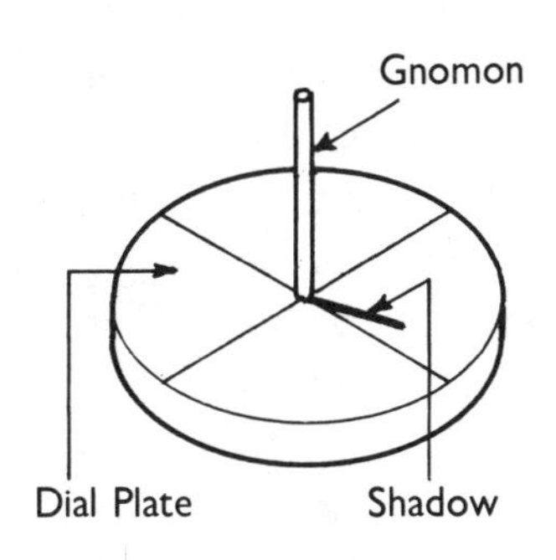

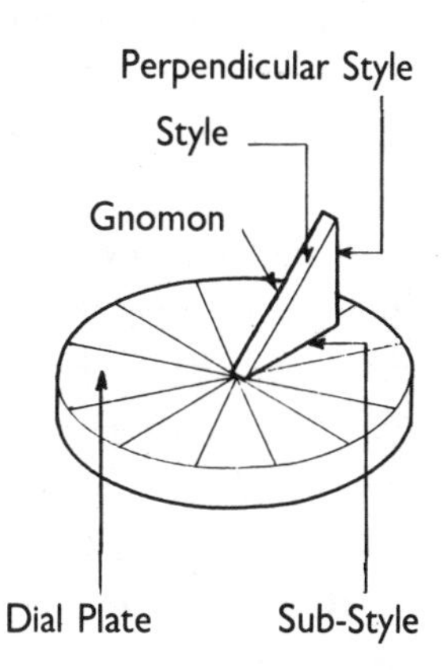

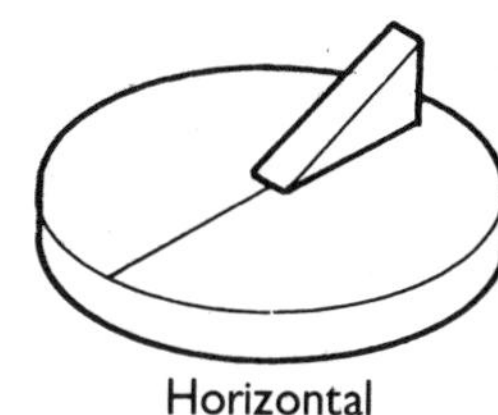
Horizontal

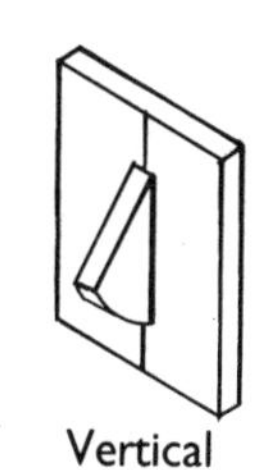
Vertical

In order to tell time, a sundial must be correctly aligned. Most sundials are so arranged that the gnomon points toward the celestial pole; that is, parallel to the earth's axis. These include both horizontal and vertical dials, cross dials, and equatorial dials. Many other types of sundial exist, such as analemmatic and polar dials, but they are more difficult to design and construct.

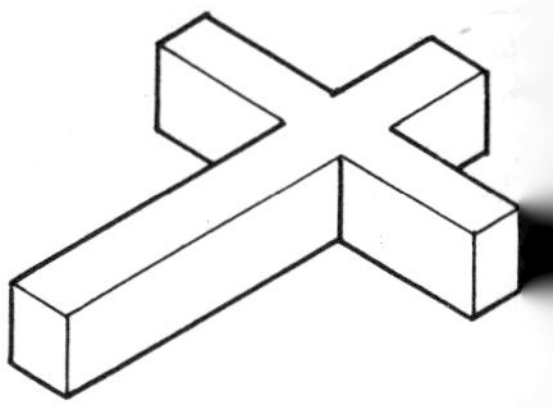

At this point, it is worth talking a bit about how positions on the earth are defined. If we draw north–south lines on a globe, we find that they come together at the north and south poles. These lines are called *meridians.* If we divide the earth's circumference at the equator into 360° and draw a meridian for each degree, we can define the east–west position of any point by saying that it is so many degrees east or west of some standard reference point. It happens that the reference point that we use is Greenwich, England. This measure of position is called *longitude.*

Equatorial

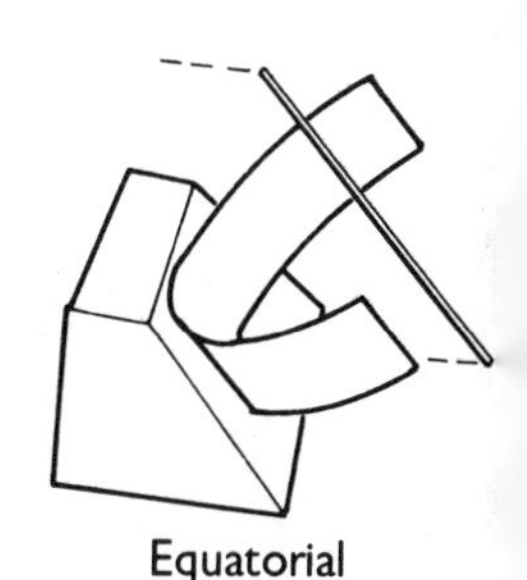
Cross

For defining north–south position, we measure how many degrees a point is north or south of the equator. If we define the equator as 0° then the north pole is 90° north. This measure is called *latitude.* By specifying both, the position of any place on the earth can be defined. As we will see, both latitude and longitude must be taken into account when setting up or reading a sundial.

A sundial tells *local apparent time* (LAT). This differs from clock time for several reasons. For one, the path of the earth around the sun is not a circle. This means that the rate at which the sun appears to move across the sky is not constant. If you were to set a clock to the time shown by a sundial on December 21 and compare the two each week, you would find the sundial reading 'fast' or 'slow' as the seasons passed. The clock is keeping *local mean time* (LMT). This is the time the sundial would show if the sun's movement across the sky were uniform. The difference between the clock and the sundial when plotted through a whole year gives us a chart called the *equation of time.* Four times a year the difference is zero. The rest of the time, the difference may be as great as 15 minutes.

The clock on your wall does not keep local mean time, but *standard time.* The trouble with both LAT and LMT is that places to the east or west of you keep a different time. This is because the earth is round. According to Albert Waugh in his book, *Sundials: Their Theory and Construction,* there is a time difference of 1/4 second in local apparent time at opposite ends of a football field if the field lies in an east–west direction. In another example, two towns only 13 1/2 miles apart in an east–west direction will differ in LAT by one minute. This sort of thing is very upsetting to railroad timetables. For their own convenience, railroads divided the country into zones. Each zone would keep the LMT of a meridian running down its center, and they were made just wide enough that there would be an hour difference between neighboring zones. This is the basis of the standard time we use today. So your LMT is not going to agree with standard time unless you happen to be on the standard meridian that runs down the center of your time zone.

How great is the difference? The sun appears to travel once around the earth, or 360°, in 24 hours. That comes to 15° of longitude per hour, or one degree per 4 minutes. Time zones are thus 15° of longitude wide, and for every degree you are east or west of a standard meridian, your LMT is a minute ahead or behind standard time. So it is possible to determine standard time with a sundial—you read the LAT, then use the equation of time to figure the LMT, and then you correct for your distance from the standard meridian. Consequently, the dial plates on sundials may be adjusted accordingly to compensate for differences in longitude; the dial is rotated four minutes of time for every degree of longitude that the dial is placed east or west of a standard time meridian.

With horizontal sundials, the angle of the gnomon is equal to the latitude and the hour line angles are also dependent on the latitude. With equatorial sundials, the dial plate must be parallel to the earth's equator. Therefore, the latitudinal position of a sundial must be accurate before a precise reading of time can be observed. Your latitude and longitude can be determined with sufficient accuracy from a U.S.G.S. topographic map or some other large-scale map of your region.

Equatorial Sundial
(with Adjustment for Latitude)

The Equatorial Sundial has both an upper and a lower dial plate; its name is derived from the fact that this double plate lies parallel to the equator. Since the sun is south of the equator during the winter months, and since the dial plate is parallel to the equator, the sun will only shine on the lower dial plate (winter dial) between September 23 and March 20. The upper plate (summer dial) tells time between March 20 and September 23. The latitude dial, of course, must be adjusted and set correctly for this to occur.

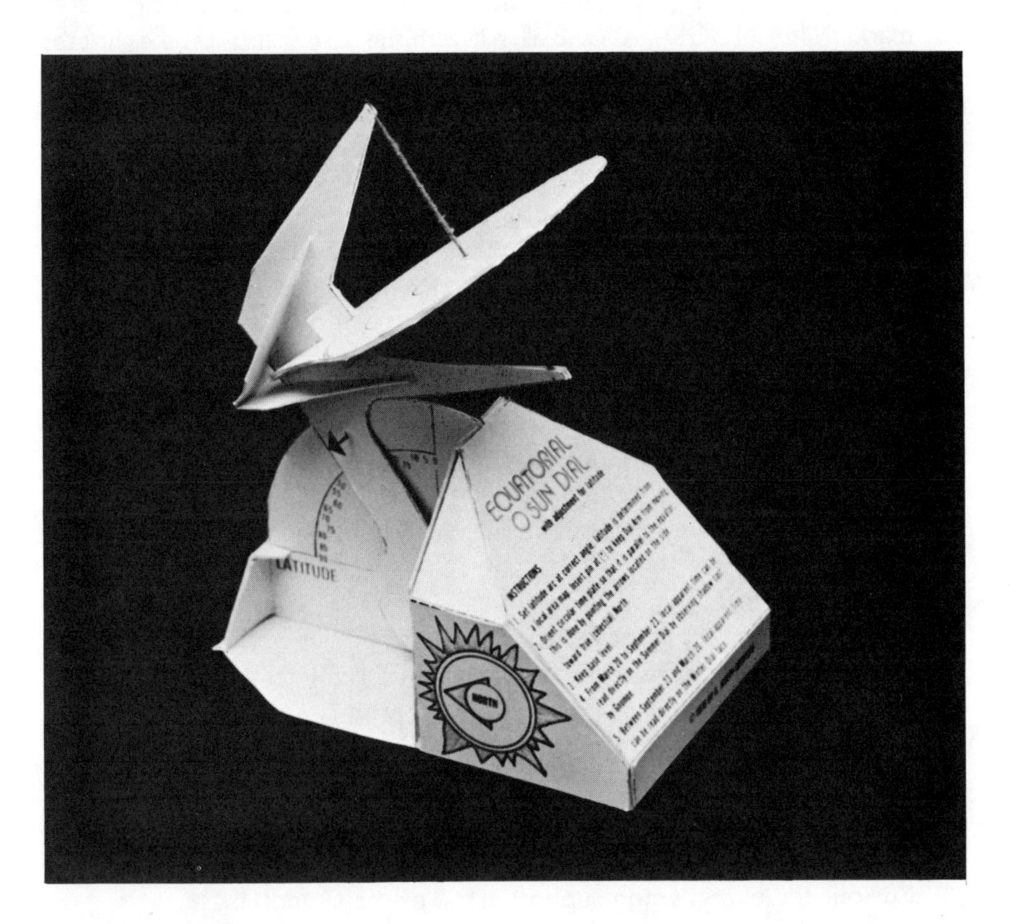

Assembly Instructions

Step A

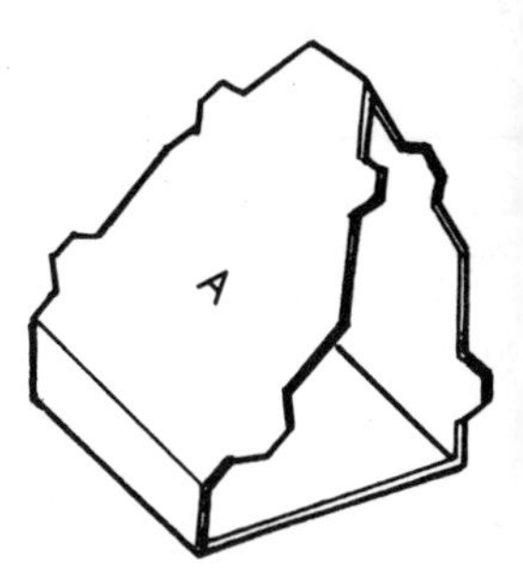

1. Cut out piece A (main body).
2. Cut out the two slots with a sharp x-acto knife or single-edged razor blade.
3. Score all dashed lines and tabs.
4. Fold the scored edges away from the cut.

Step B

1. Cut out piece B (base).
2. Cut out the two notched slots.
3. Carefully cut out the eight slots used in conjunction with the glue tabs from piece A; use a sharp x-acto knife.
4. Gently score and fold all dashed lines.
5. Spread some white glue on the middle rectangle of the main body (the area marked "MAIN BODY A") and align it with the base B; insert the eight tabs from piece A into the matching cutout slots of the base (B).

Step C

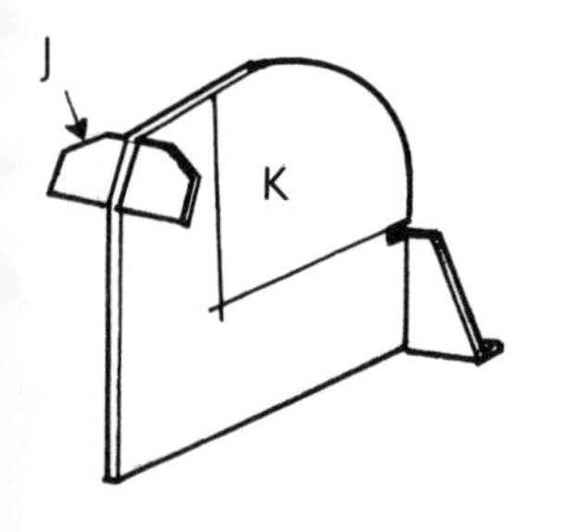

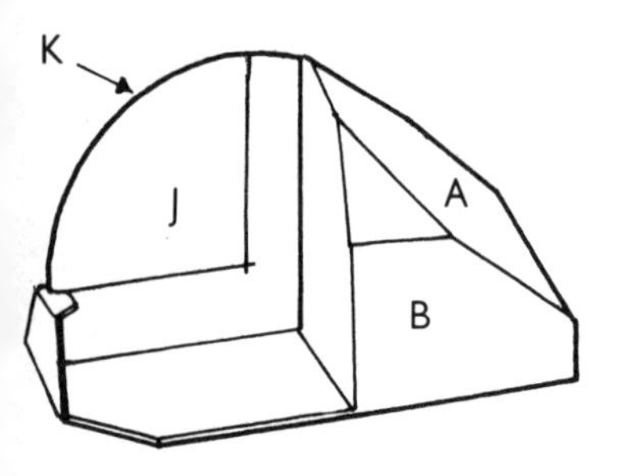

1. Cut out both latitude scales (pieces J and K).
2. Cut out the notch on each scale; the notch is located on the 90° latitude line.
3. Punch out the hole on each scale (as indicated).
4. Score the two valley folds on each scale from the reverse side and fold toward the dotted line (printed side).
5. Gently score the remaining dashed line on each scale and fold away from the line.
6. Attach the two latitude scales to the base by placing both scales together then inserting the edge marked with a centerline into the cutout slot of the main body. Position all tabs into their appropriate cutout slots and glue in place.

Step D

1. Cut out the two gnomon support arms, pieces G and H.
2. Carefully cut out the three slots of each piece.
3. Cut out the two circular reinforcements, pieces E and F.
4. Punch out the holes marked (+) on both the two gnomon support arms and the two reinforcement pieces.
5. Using a straight pin, punch a pinhole in each gnomon support arm at the denoted spot (+).
6. Carefully slit notch for gnomon string on each arm.

Step E

1. Cut out the upper dial (piece C) and the lower dial (piece D).
2. Cut out the two slots on each piece.
3. Insert a pushpin through the exact center of each circular dial face. The gnomon string or straight metal rod will later be inserted through this hole.
4. Cut out the two small dial support tabs (M); these help stiffen and align the dial plates.
5. Cut out the remaining support arm reinforcement (piece L) and cut out the two slots with an x-acto knife.
6. Score the remaining ridge fold and then gently fold away from the cut. NOTE: Do not score the light centerline that is perpendicular to the ridge fold.
7. Place some white glue or rubber cement on the back side of both the upper dial and lower dial and then place them together, forming one solid piece with printing on both sides.

Step F

1. Insert the double dial plate into the long slot of the gnomon support arm. The dial plate should now be located above the "INSTRUCTIONS" of the main body.
2. Insert one dial support tab (M) on each side of the gnomon support arm; these are inserted into slots cut out of the dial plate.
3. Insert the support arm reinforcement (piece L) into the remaining two slots of the gnomon support arm.

Step G

1. Place some white glue or rubber cement on the back side of the two gnomon support arms on the upper triangular support only (area with slots) and align together, gluing as one piece.
2. Slightly separate the lower circular discs and adjacent arm (the part with an arrow); place each arm over the latitude scale so that both the holes punched in the circular discs and those at the center point of the latitude scale line up. NOTE: The arrow on each arm must face away from the "INSTRUCTIONS", enabling the dials to face south.

3. Place the two reinforcement pieces E and F over each circle of the lower gnomon support arm and insert a small paper fastener through the aligned hole.

4. Stretch a piece of heavy thread or string (or straight piece of coat hanger or welding rod) through the center hole of the dial plate and attach it to each end of the support arm; this is the GNOMON.

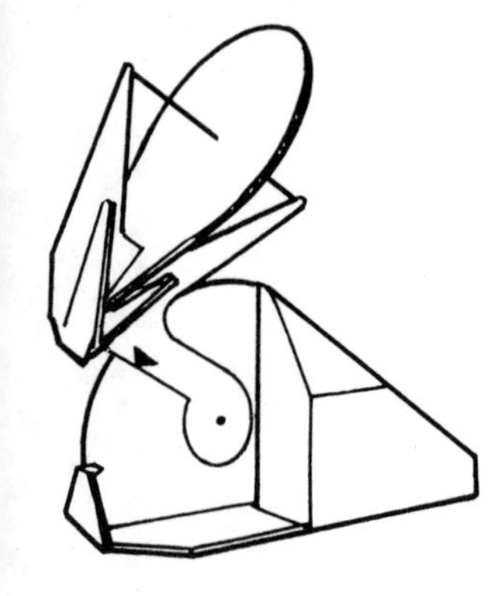

Operating Instructions

1. Set the correct latitude on the latitude scale and insert a pin into the (+) mark near the arrow; this will prevent the latitude adjustment from shifting.

2. Keep the base level.

3. Orient the NORTH arrow toward true (celestial) north.

4. From March 20 to September 23, local apparent time can be read directly on the upper summer dial by observing the thin shadow cast by the gnomon.

5. Between September 23 and March 20, local apparent time can be read on the lower winter dial.

6. To convert from Standard Time to Daylight Savings Time, add one hour.

Bowstring Equatorial Sundial
(with Adjustment for Latitude)

The Bowstring Equatorial Sundial provides a dial plate that enables the reader to face the instrument directly and read the time, regardless of the season or month. When the sundial is aligned toward true north the gnomon will cast a thin shadow, indicating the time of day. If the latitude dial is set accurately the gnomon will be parallel to the earth's axis. Therefore, each hour line will measure 15° of rotation as the earth spins on its axis.

If the reader would like to make a longitude adjustment to either equatorial sundial, the dial plate must be rotated four minutes for every degree of longitude which the local longitude differs from that of the major time zone in your region. The standard meridian for each of the major time zones in the United States are: Atlantic, 60°W; Eastern, 75°W; Central, 90°W; Mountain, 105°W; Pacific, 120°W; Yukon, 135°W; and Alaska-Hawaii, 150°W.

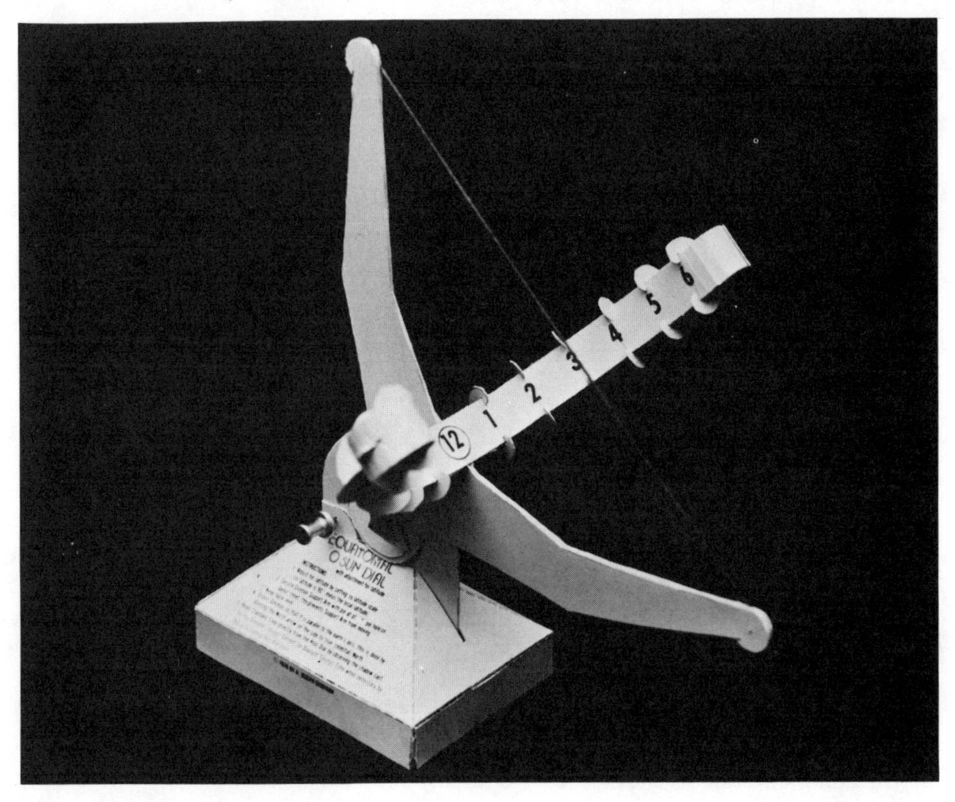

Assembly Instructions

Step A

1. Cut out piece A and piece B (gnomon support arms).

2. Cut out window on piece B.

3. Score glue tabs A and B (both pieces) on underside and fold forward.

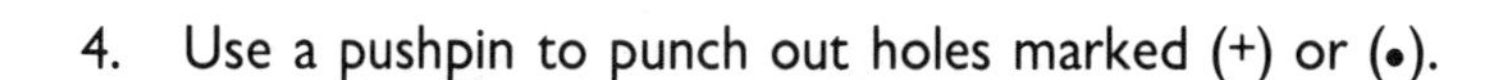

4. Use a pushpin to punch out holes marked (+) or (•).

Step B

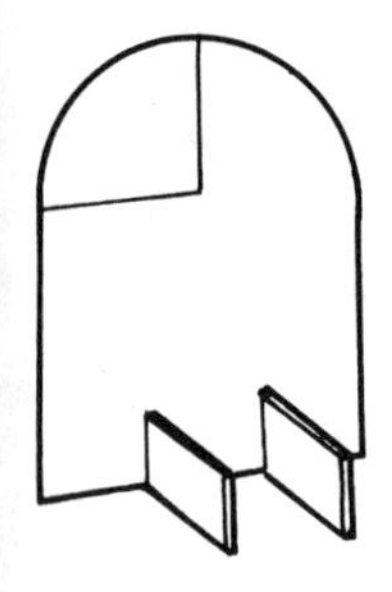

1. Cut out co-latitude dial (piece D and piece E).
2. Cut out slots C and D.
3. Align piece D and piece E; punch out center hole marked (+).
4. Cut out co-latitude dial support pieces (L), insert them into slots at the base of the co-latitude dial (pieces D and E) and securely tape them together.

Step C

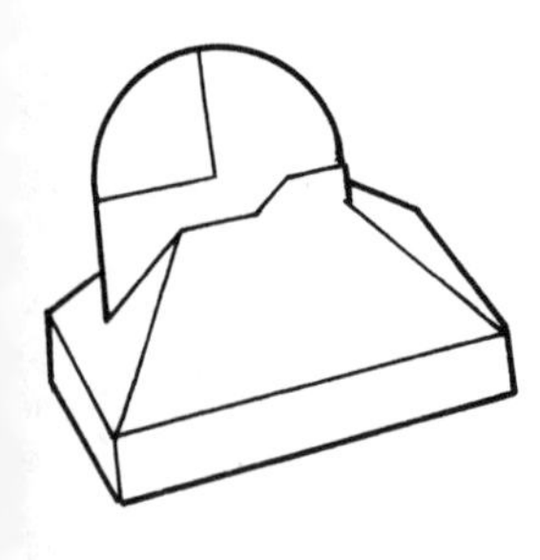

1. Cut out equatorial sundial base (piece M).
2. Cut out all slots with an x-acto knife.
3. Score and bend all ridge folds.
4. Before inserting and gluing tabs, place the co-latitude dial into the cut slot. Be sure that the dial is located on the side next to the north arrow (east side) and that it rests firmly on the inside bottom of the sundial base.
5. Insert glue tabs into the cutout slots and glue securely with white glue.

Step D

1. Carefully cut out the hour support arm (piece J).
2. Cut out slots with a sharp x-acto knife.
3. Score and fold all ridge lines.
4. Cut out all twelve dial rib supports (piece N).
5. Carefully cut out all slots with a sharp x-acto knife.
6. Cut out the hour dial (piece C).
7. Carefully cut out the two slots.
8. Lightly score the solid lines from each line adjacent to the "6" on both ends; continue until all are scored. This is necessary so that the hour dial can be placed around the smooth curve at each end of the hour dial support arm.
9. Slide the twelve hour dial supports (piece N) onto the hour dial (piece C).

10. Next, place the hour dial (piece C) onto the hour dial support arm (piece J) by inserting each of the rib supports into the cutout slots.

11. Place the cutout slot of each curved end of the hour dial over the tab of the hour dial support arm.

12. Insert the two hour dial interlocks (piece P) into the two cutout slots located under the support arm tabs and glue securely with white glue.

Step E

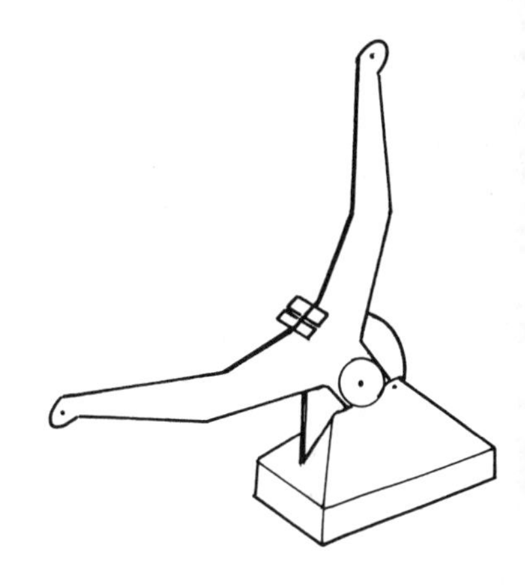

1. Position both gnomon support arms (pieces A and B) on the co-latitude dial. Place the support arm reinforcement (piece G) on the dial side of the support arm (piece B) together with one reinforcement, piece H. Next, place the other support arm reinforcement (piece F) on the other side (piece A) together with optional reinforcements, pieces H and F. Insert a paper fastener through all the aligned holes.

2. Slide the completed hour dial support arm assembly onto the gnomon support arm (piece A) by aligning the cutout slots of each piece.

3. Glue the hour dial to the four glue tabs (A and B); this secures the hour dial to the support arm. Next, glue tab X of the hour dial support arm to tab X of the support arm reinforcement (piece G).

4. Glue tab Y of the hour dial support arm to tab Y of the support arm reinforcement (piece F).

5. Carefully stretch a black thread or string between and through each hole at opposite ends of the gnomon support arm and insert it into the cutout notch near the hole. Once the string is tight and straight, glue it in place; this is the GNOMON of the sundial. This completes the assembly instructions.

Operating Instructions

1. Adjust the sundial for latitude by setting the co-latitude scale on the correct mark (co-latitude is 90° minus the local latitude).

2. Secure the gnomon support arm with a straight pin at the point marked (+); this prevents the support arm from moving.

3. Keep the base level.

4. Point the north arrow on the side toward true north; this orients the gnomon so that it is parallel to the earth's axis.

5. Read the correct time directly from the hour dial by observing the shadow cast by the gnomon. Compensate for Daylight Savings Time when necessary by adding or subtracting one hour.

Cross Sundial
(with Adjustment for Latitude)

Cross sundials have been used for several hundred years; they are usually designed as memorials for cemeteries, city parks and college campuses. The edges of the cross serve as gnomons, casting shadows onto an adjacent dial plate marked on the side. The edge of each shadow will appear as a straight line, marking the time of day as referenced by the dial plate scale. Both the latitude adjustment dial and the north arrow must be positioned accurately before the time will be correct.

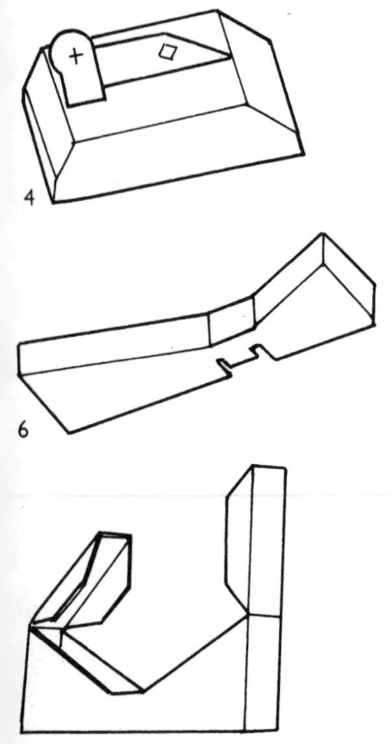

Assembly Instructions

1. Cut out piece A (base).
2. Score and bend valley folds.
3. Score and bend ridge folds.
4. Assemble piece A by joining tabs with glue.
5. Cut out piece B (sundial N–S body).
6. Score and bend valley and ridge folds, then assemble with glue.
7. Cut out piece C (sundial E–W body). Score, fold, then glue all pieces together.
8. Cut out piece D, latitude dial.

9. Attach sundial E–W body (piece C) to N–S body; this is done by aligning the two edges between the 11 o'clock and 1 o'clock markings of piece C with the notches on piece B located between the 5 o'clock and 7 o'clock markings and sliding the two pieces together. Be absolutely sure that the dial markings on piece C face toward the short end of piece B.

10. Attach the latitude dial (piece D) to the bottom of the sundial cross by inserting the appropriate tabs into their respective slots (M to M and N to N). When the dial has been inserted, slide it forward, locking the tabs in place.

11. Attach the sundial cross to the base by inserting the latitude dial into the latitude dial support. The latitude marks should be visible in the window. Insert a pin through the pivot arm located on each side of the dial.

12. Adjust the latitude dial for the latitude of your specific location, then insert a short pin through the (+) mark on the dial to prevent the dial from moving. Orient the north arrow toward true (celestial) north and adjust the base so that it is as level as possible. The time can now be read by observing the shadow line which has been cast onto the dial face.

NOTE: To correct for Daylight Savings Time, add (or subtract) one hour to the time indicated on the dial face.

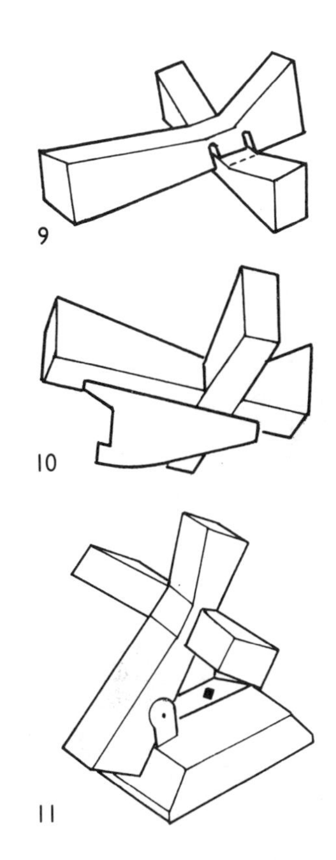

PART THREE

Solar Collection Devices

Flat-Plate Collectors

When you close the windows of a car and face the front windshield toward the sun on a bright, clear day, the inside of the car heats up rapidly. This phenomenon is known as the *greenhouse effect;* it is the basic principle underlying the design of flat-plate collectors. When the sun's short wave rays enter an enclosed box through a glass or plastic cover and strike something dark or black, long wave heat rays are sent back. However, the flat, clear glass only lets the short sun rays pass through and prevents most of the long heat waves from leaving. This allows the box to become a collector of heat. The name flat-plate collector is used to describe it because its dimensions are usually wide and shallow.

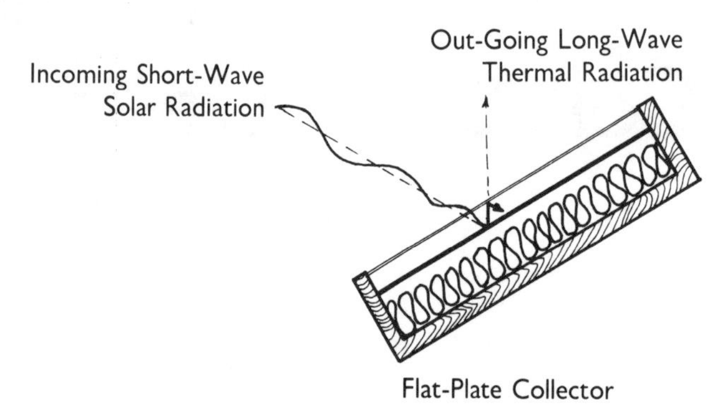

Flat-Plate Collector

The Romans during Pliny's time were aware of the ability of transparent window coverings to trap heat. Glass and mica were both used to heat sun rooms in houses (called the heliocaminus) and for greenhouses to keep plants warm in the winter. They were also used, in part, to heat the public Roman baths. True research did not begin, however, until 1767 when the French–Swiss scientist Horace de Saussure began to experiment with glass as a means of trapping heat derived from the sun. He constructed a small wooden box and lined it with black cork; he then covered it with three layers of glass to prevent heat loss from occurring through the top. With continued experimentation he reached a temperature of 230° F inside the box. This was the forerunner of the present flat-plate collector.

Numerous designs exist for flat-plate collectors; many different types of materials are used in their construction. However, most have three components that are housed in an enclosing framework; these are:

1. Transparent cover
2. Absorbent surface
3. Heat transfer fluid

The transparent cover is usually glass or plastic; it must allow the sun's rays to pass through but prevent most of the heat waves emitted by the black or dark absorbent surface from leaving. It also protects the materials housed inside the box from the weather and lessens the amount of heat loss caused by convection (air movement).

The absorbent surface is usually some type of dark or black metal plate that absorbs the short wave solar radiation and converts it into long heat waves; it then slowly re-radiates the heat back into the box.

The heat transfer fluid is usually either air or water. It transfers the heat from the box to other parts of the building and into a storage medium. Since either air or water can be used, *air-type collectors* and *water-type collectors* are terms commonly used to describe each type.

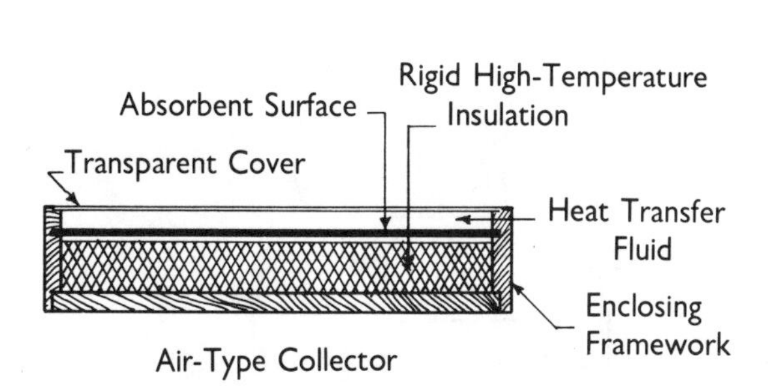

Air-Type Collector

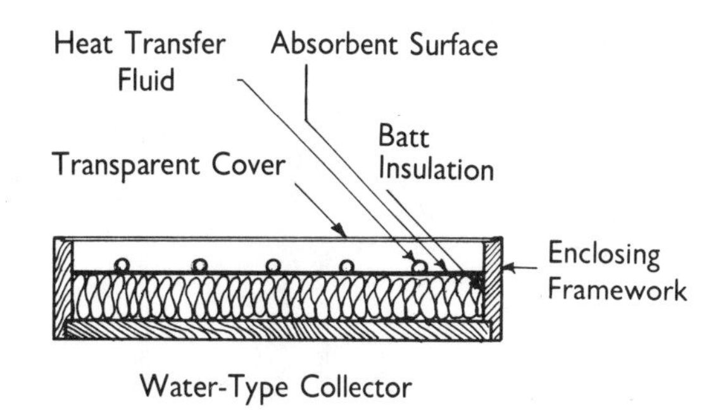

Water-Type Collector

Once the inside of the collector has been heated, the heat is either used immediately or stored for later use. For storage, water is usually circulated through large storage containers of water to transfer its heat and air is circulated through large volumes of clean rock. These materials are then insulated to prevent the heat from escaping before it can be used.

Air-type collectors are usually more trouble-free, lighter and easier to construct than water-type collectors, and they can operate at lower temperatures since the hot air can be circulated immediately for space heating needs. Another advantage is that air systems can be added directly to an existing forced air heating system. The two big disadvantages are the larger space required for the rock storage and also the larger ductwork required, especially when trying to fit large ducts into existing walls and ceilings of older houses.

Water-type collectors are often used for domestic hot water heating because of the easy adaptability of a domestic solar water system to an existing hot water tank. Domestic hot water systems account for the greatest use of solar energy in the world today.

Active and Passive Solar Heating Systems

Active heating systems use mechanical (electrical) components such as pumps, fans, thermostats and valves to transport and regulate heat gathered by the collectors and to distribute it throughout the building. This type of system normally consists of six components working together to provide space heating for a building. These components are:

A. Flat-plate collectors (water or air)

B. Distribution system (water pipes and air ducts)

C. Heat storage (water or rocks)

D. Auxiliary (backup) heat source

E. Mechanical controls (water valves, pumps and fans)

F. Heat exchanger (transfers heat from storage into the building)

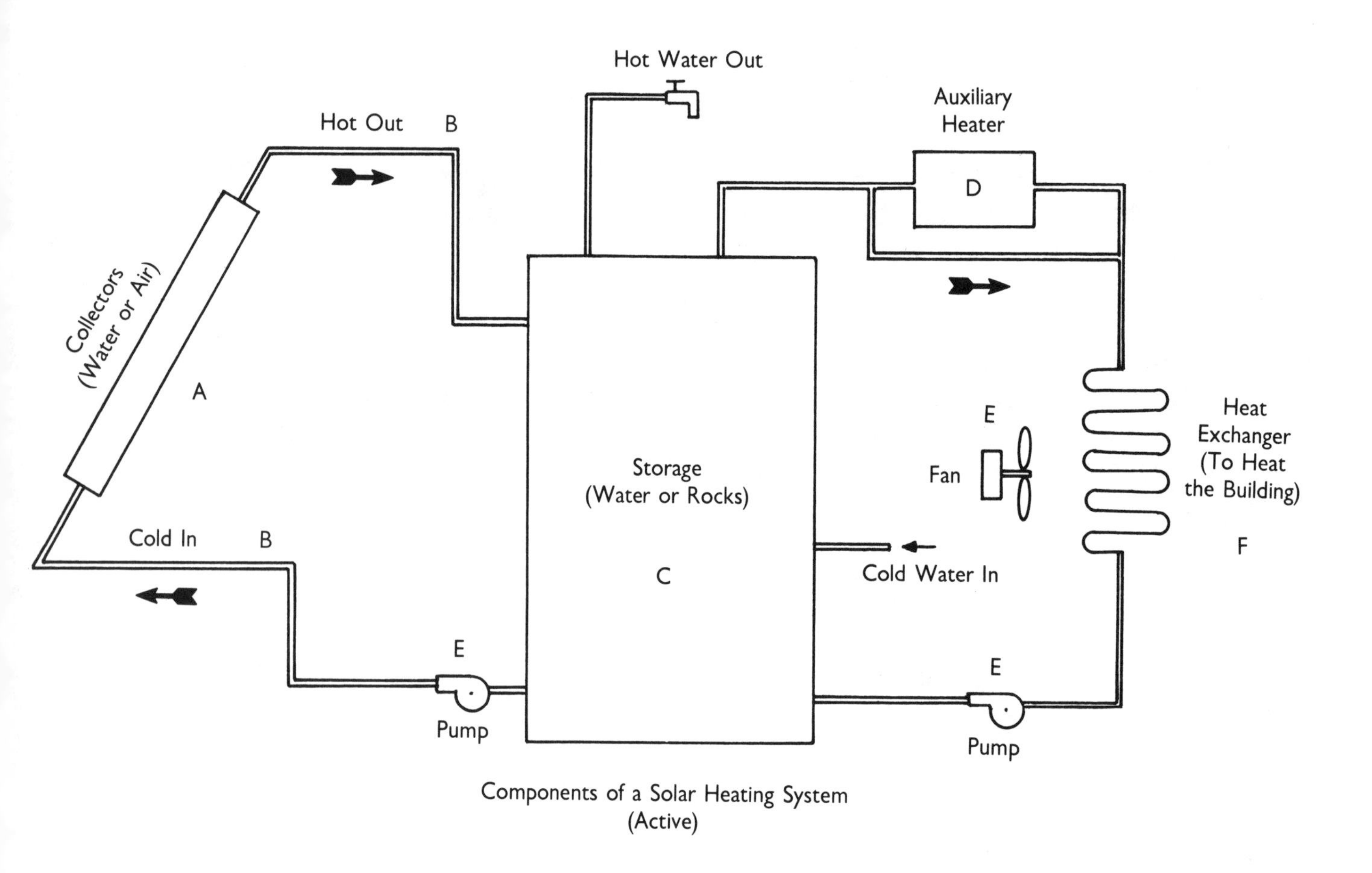

Components of a Solar Heating System
(Active)

Passive heating systems rely on natural radiation, conduction and convection currents to transfer heat without using electrical controls or motors. The structure itself is designed to collect and store the heat; south-facing attached solar greenhouses and large south-facing windows are used to collect the sunlight. A carefully placed storage mass composed of thick masonry walls, floors, and water-filled containers of a dark color are positioned to absorb the solar radiation and store heat, slowly releasing it throughout the cooler night. The windows are insulated at night to lessen the heat loss from the interior.

Both active and passive concepts can be used in the same building. The most cost-effective active system is one that heats hot water for domestic household uses; passive systems are ideally suited for space heating.

Model Solar House

The Model Solar House is intended to be an educational model from which the concepts of solar heating can be studied. It uses both active and passive solar heating systems: an active domestic hot water system with collectors on the roof and a passive solar greenhouse with large south-facing glass. The solar greenhouse is attached to the living room to provide additional space heating for the house. A 12" thick masonry Trombe wall separates the greenhouse from the living room; a Trombe wall is a storage wall that collects heat, slowly releasing it throughout the colder night.

The walls are constructed of 2" by 6" studs rather than the more conventional 2" by 4" studs commonly used in house construction; thicker insulation can then be used to cut down on heat loss. A space has been provided for solar hot water storage near the washer and dryer (and lower bathroom); this is the small area marked "HW" on the Main Floor Plan. Most windows in the house are small to further cut down on heat loss; it is assumed that the large windows in the living room will have thermal insulating shades. The high clerestory window can also be used for solar gain if the wall behind it has sufficient thermal mass (storage capacity) to retain the heat overnight. A wood stove has been placed near the Trombe wall so that the excess radiant heat from the back side will heat it to supplement the heat gain from the solar greenhouse.

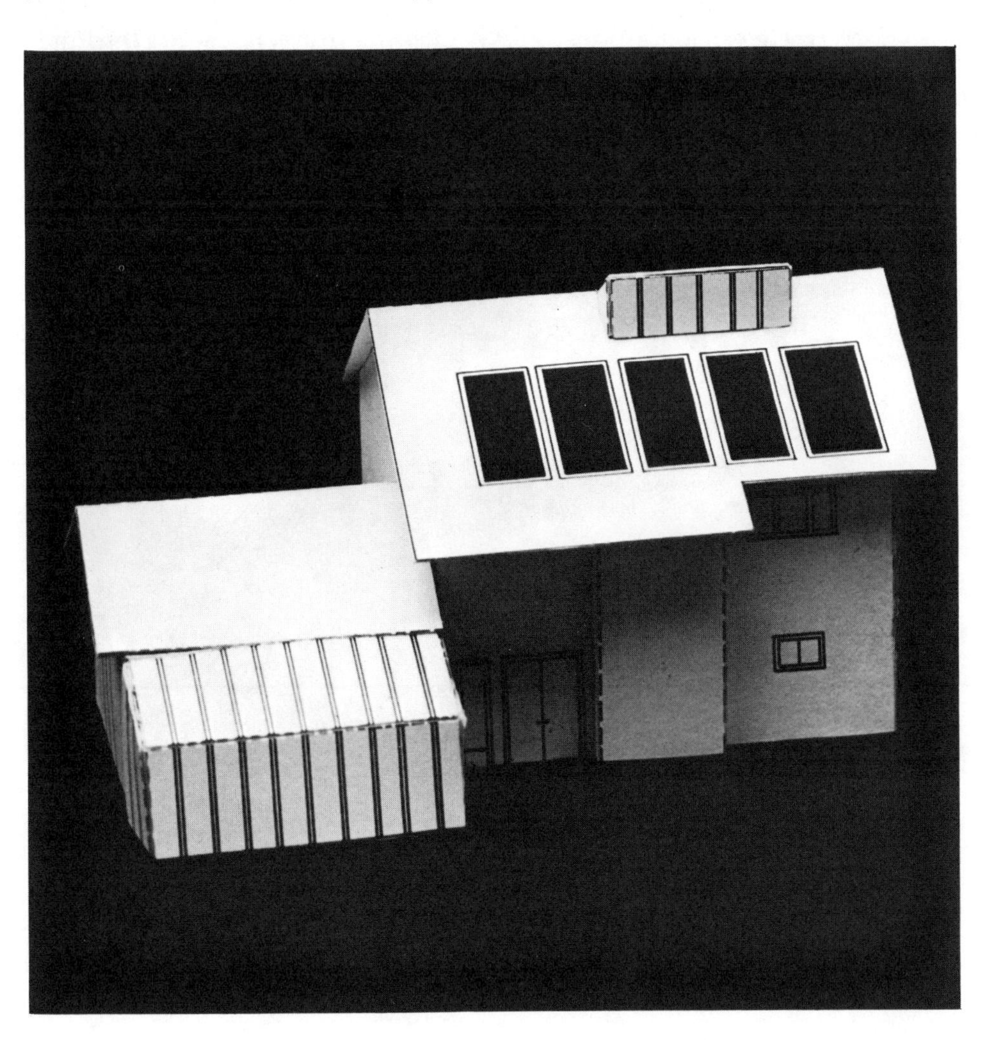

Assembly Instructions

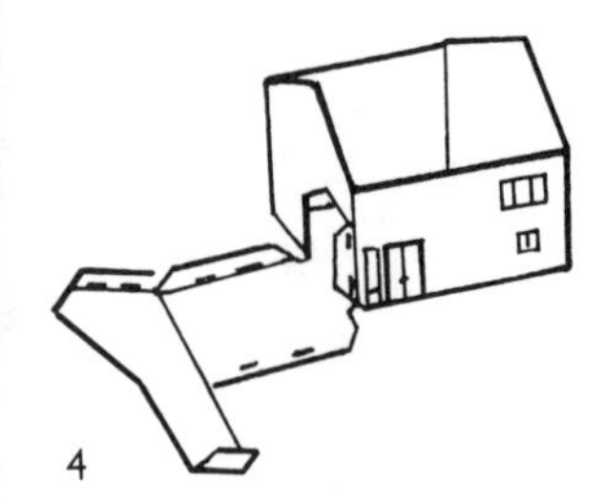
4

5

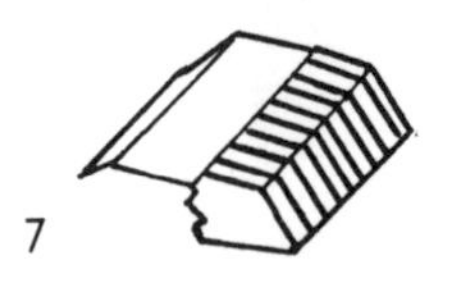
7

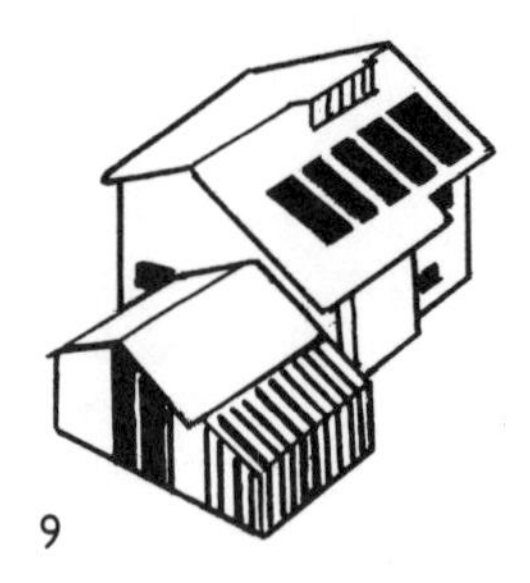
9

1. Cut out all pieces on solid lines.
2. Cut out all notches.
3. Score ridge folds.
4. Assemble and glue the exterior walls (pieces A1, A2, A3); insert numbered tabs into their corresponding slots. Insert and glue the remaining tabs into their slots.
5. Attach and glue the clerestory window (piece F) to the main roof (piece D). Notice the solid line that must be cut out above the black solar collectors.
6. Attach the main roof (piece D); apply glue to the upper edges of the exterior walls first, however, for a permanent bond.
7. Assemble and glue the greenhouse with attached roof (piece E).
8. Attach and glue piece E (greenhouse) to the main house; see the floor plan for correct greenhouse placement. Attach the rear wall (piece H) to the north side of the living room.
9. Attach and glue the front wall protuberance (piece G).

The model is now complete; study the lower and upper floor plans for the circulation patterns and general layout. The solar greenhouse and rooftop solar collectors must face south for maximum solar heat gain.

Solar Oven

In 1837 an English astronomer, John Herschel, constructed a simple solar oven designed for cooking meals on his expedition to the Cape of Good Hope. He built a black box with a double-layer glass cover and buried it in the sand for insulation. A temperature of 116° C (240° F) was recorded.

The solar oven in this book uses side panels to reflect extra light from the sun into the box through the transparent cover; this increases the inside temperature. The greenhouse effect traps heat inside, behind the transparent cover. Once the oven box is insulated and the maximum temperature has been reached, the window can be covered with an insulating pad to retain the heat for longer cooking time. Higher temperatures will be reached if the solar oven is doubled in size and well-insulated; follow the examples given in the chapter entitled "Hints for Making Projects" for an easy method to do this.

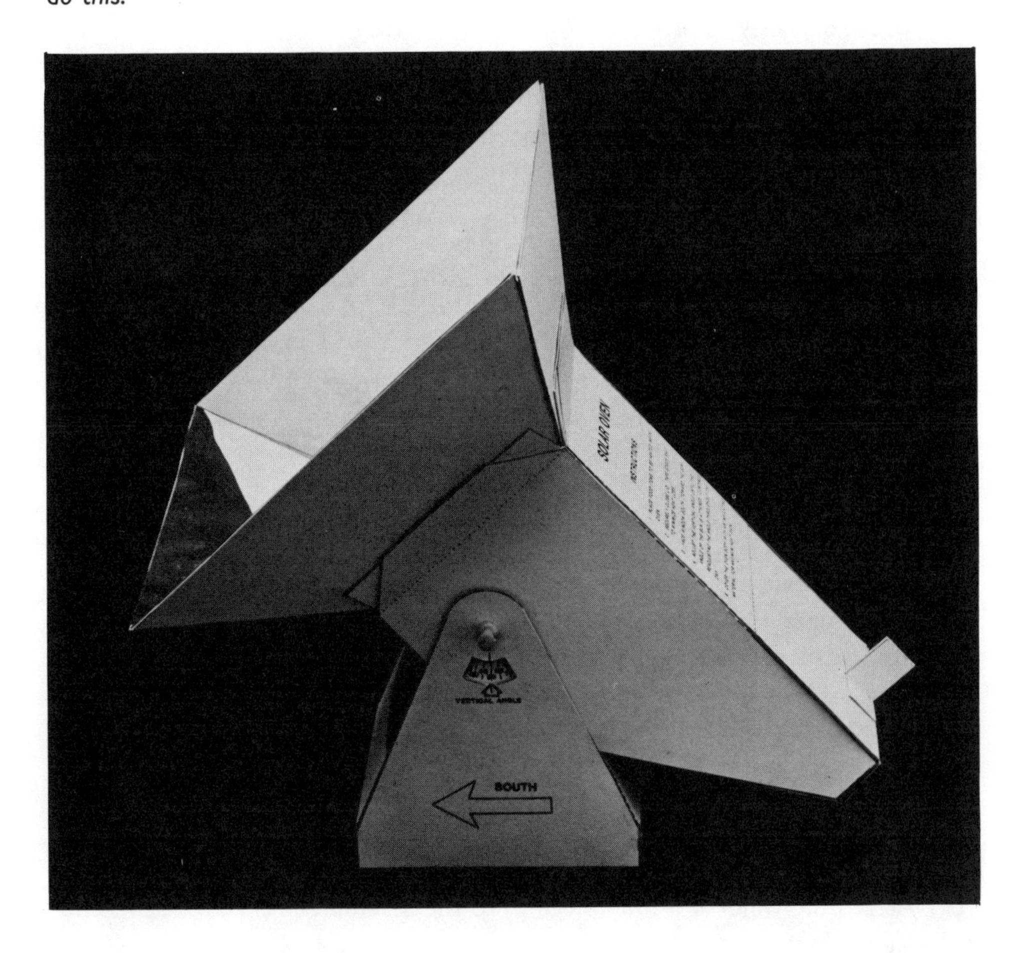

If the oven does not heat up properly, any one of the following reasons may be the cause:

1. *It is not focused properly on the sun.*
2. *Clouds or smog may obscure solar radiation.*
3. *The wrong type of clear transparent cover may have been used.*

4. *Too much heat loss is occurring; tape all cracks and insulate the box.*
5. *A shiny cooking container reflects light. Use a dark-colored container to cook the food.*

For best results, use the Sun Angle Viewer to orient the solar oven toward the sun.

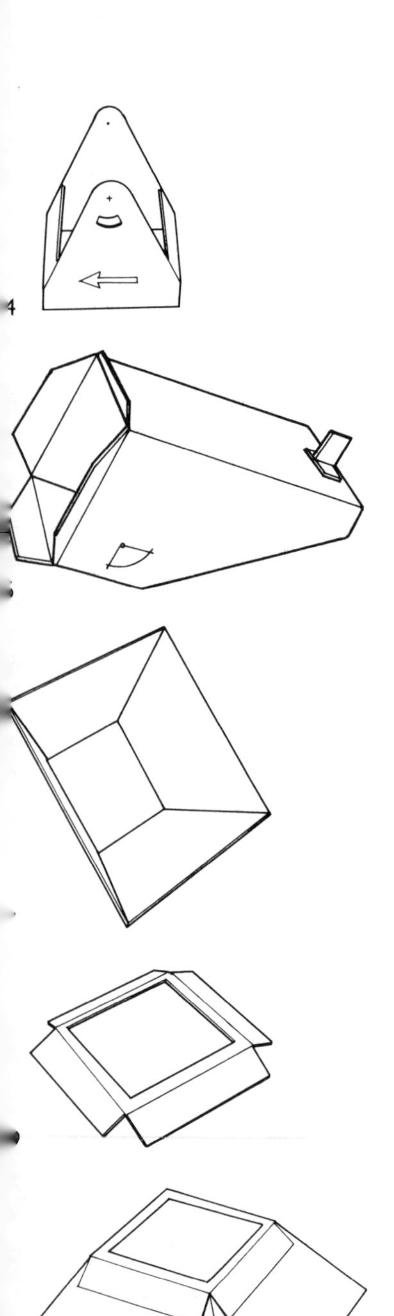

Assembly Instructions

1. Cut out all pieces on solid lines.
2. Score all ridge folds.
3. Score all valley folds.
4. Assemble the base (piece G) and glue securely; punch a small hole at each of the two (+) marks. A straightened coat hanger will be inserted through these marks later to provide an axis of rotation for the solar oven.
5. Assemble and securely glue the main body (pieces A and B) together. All seams must be airtight except for the access lid, which must operate freely. Glue the finger-hold (piece H) to the top of the lid for better access.
6. Center and glue the upper lid clasp (piece J) to the lower part of the lid; glue the lower lid clasp (piece K) to the main body directly below the upper lid clasp. Be sure both holes are closely aligned.
7. Assemble the reflective side panels (pieces C and D).
8. Cut out four pieces of aluminized mylar or aluminum foil using the template (piece F). Using a spray adhesive, attach the reflective mylar to the inside of the reflective side panels.
9. Cut out a $4\frac{1}{2}''$ by $4\frac{1}{2}''$ piece of clear mylar or clear self-basting oven bag and glue it to the window frame (piece E). Be sure all edges are glued airtight and that the bag or mylar has been stretched as tightly as possible, with no wrinkles.
10. Attach the window frame (piece E) to the assembled reflective panel and glue tightly.
11. Attach and glue the main body to the reflective panel; all seams must be airtight.
12. Attach the completed solar oven to the base by aligning the angle adjustment marks with the angle adjustment window. Insert a straight metal rod (a straightened metal coat hanger or welding rod) through the four (+)

marks; this is the axis. Two paper fasteners, one on each side, work equally well if inserted from the oven interior.

This completes the assembly instructions.

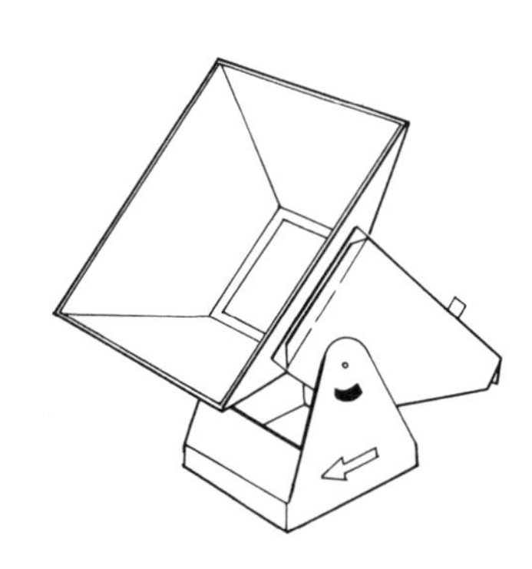

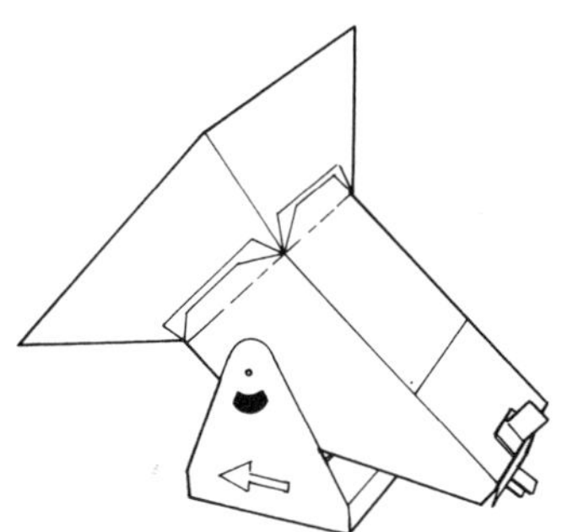

Operating Instructions

1. Place food items to be heated inside a dark or black metal container, then place it inside the solar oven. Try toasting a cheese sandwich to start with.
2. Close the lid and seal the edges shut with tape to minimize heat loss.
3. Face the window south (toward the sun).
4. Adjust the vertical angle until the angle of the sun is attained. Continue re-adjusting the angle throughout the day.
5. Cover the oven body with an insulating material for maximum heat gain.
6. Insert a paper fastener through the upper and lower lid clasps to keep the lid tight.
7. Use an oven thermometer, if possible, to observe the interior temperature.
8. When the desired temperature (or maximum temperature) is reached, cover the transparent window with an insulating pad to retain the interior heat.

PROJECT
7

Solar Food Dehydrator

For centuries the best way to preserve food has been to dry it; dried dates, figs and grapes (raisins and currants) have been popular since ancient times. The North American Indians dried meat and fish in the sun to preserve them for future use, and soldiers consumed dried food during the Civil War.

Dehydration is the process of preserving foods by removing the water from them. Bacteria need moisture to grow; by reducing the moisture content, their growth is inhibited. To effectively dry food in the sun, three components are needed. These are:

1. *Heat*
2. *Free circulation of air*
3. *Protection from insects*

The food dehydrator uses reflective side panels to concentrate the sun's rays inside the box to generate heat; it is essentially a flat-plate collector with a transparent cover, black absorber plate and a storage rack inside to hold fruit or vegetables for drying. With a small breeze, air circulates through the front and back vents to eliminate moisture inside the box. The vent holes are small to inhibit insect infestation; they may have to be enlarged later and covered with cheesecloth for protection against insects if no air movement is occurring. The main body should be insulated to prevent heat loss if the required 135° to 150° F temperature is not reached.

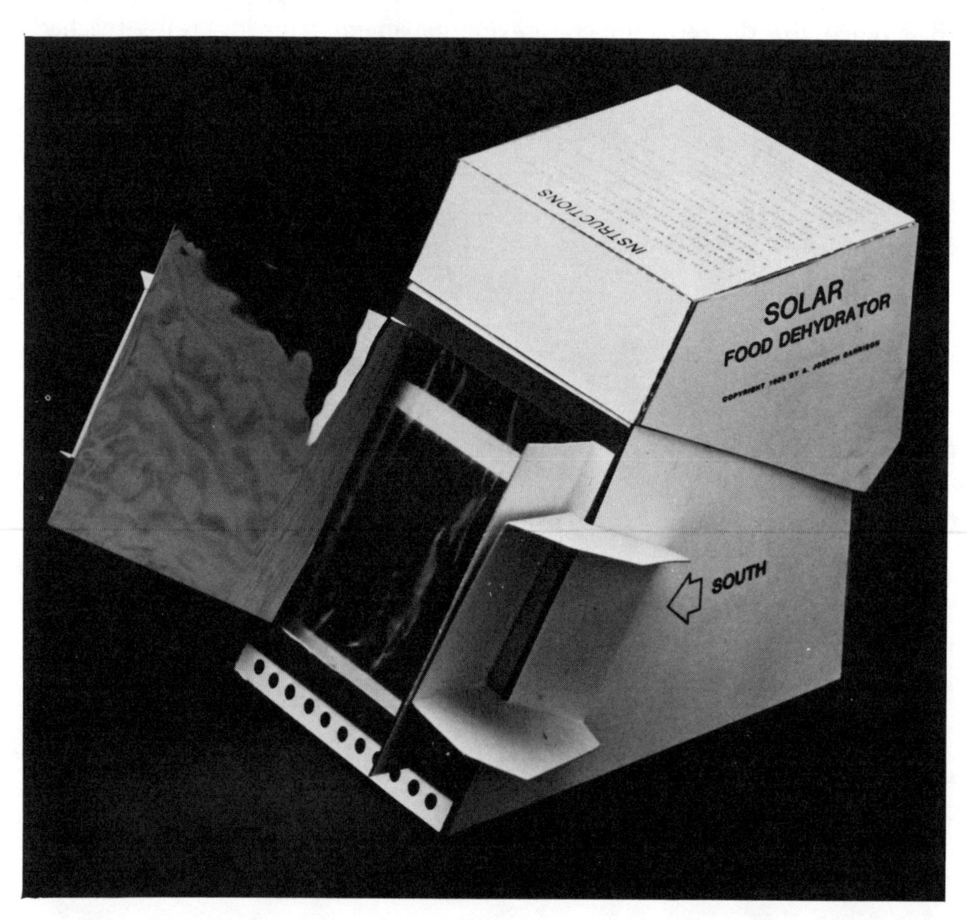

The food dehydrator should be used on a bright, sunny day with a slight breeze blowing. Sometimes the food is blanched before drying to loosen the fibers, remove bad odors and obtain more uniform evaporation of moisture in the drying process. A cookbook should be consulted for complete instructions and blanching times for vegetables. Some basic guidelines used for solar food dehydration are as follows:

1. *Keep the food dehydrator pointed toward the sun at all times.*
2. *Keep the food items separated while they are inside the dehydrator; this allows air to circulate freely around the food.*
3. *Wash and slice the fruits and vegetables into thin strips for best results.*
4. *Turn the drying food at least twice a day.*
5. *Take the food inside at night or during cloudy weather.*
6. *Do not exceed temperatures of between 135° and 150° F for extended periods of time.*
7. *If insects are getting inside the dehydrator, enlarge the vent holes and cover them with cheesecloth.*
8. *As a precautionary measure to prevent insect eggs from being hatched, the dried food can be heated to 180° F for 5 to 10 minutes.*

Assembly Instructions

1. Cut out all pieces on solid lines.
2. Punch holes marked (o) with a large paper punch.
3. Score ridge folds.
4. Assemble and glue the base (pieces A, C, H and J). Do not attach the back (piece L) at this point.
5. Cut out a 6" by 5½" piece of clear mylar or clear self-basting oven bag and glue it over the front window (piece A). Be sure all edges are glued firmly and that the mylar or oven bag has been stretched as tightly as possible, with no wrinkles. Be careful not to cover the front vent holes with the plastic.
6. Cover the window with the front frame (piece B).
7. Assemble and glue the top cover (pieces D, E, F and G). Piece G goes inside the back cover, forming a large vent in the back of the lid.
8. Assemble and glue the interior frame supports (pieces T, U and S). Cut out 10 straightened coat hangers to length (5⅞"), clean them with steel wool,

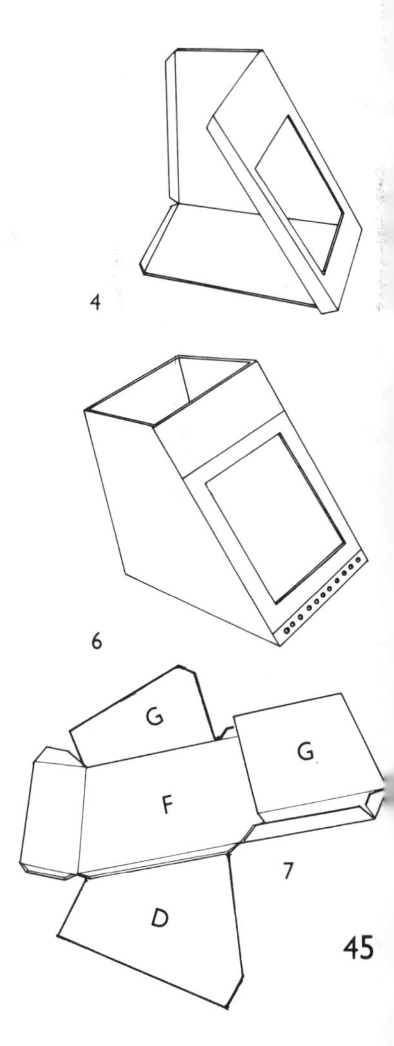

then lay them across the "V" grooves notched into pieces T and S. These metal rods hold the food to be dried.

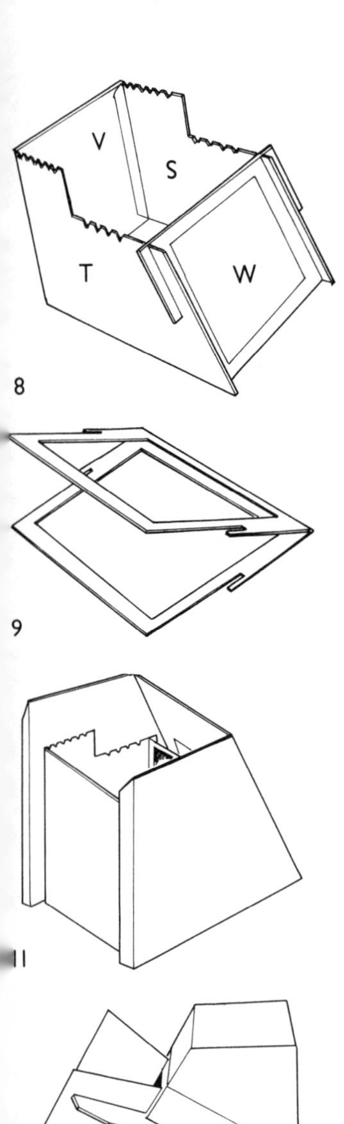

9. Cut out the frame for absorbent surface (piece W); cover the window with between 4 to 10 layers of aluminum foil that has been blackened with flat black spray paint.

10. Place the interior support frame inside the food dehydrator then attach piece W to it by aligning notch X with notch X and notch Y with notch Y.

11. Glue the back (piece L) to the main body.

12. Cover the reflective side fins (pieces M and N) with reflective mylar or aluminum foil then glue them to the front panel of the main body. Piece M goes to the right (west) side of the window and piece N goes to the left (east) side of the window.

13. Attach the two side fin gussets (pieces P and Q) to the main body; glue them securely with white glue.

The Solar Food Dehydrator is now ready for use; follow instructions printed on the top cover.

Focusing Collectors: Concentrating the Sun's Energy

Concentrating the sun's rays to get high temperatures for practical purposes is not a new idea. The first known use involved the Greek mathematician and physicist Archimedes in 212 B.C.: he reputedly set the attacking Roman fleet of Marcellus on fire by means of a "burning glass composed of small square mirrors moving every way upon hinges which when placed in the sun's direct rays directed them upon the Roman fleet so as to reduce it to ashes at the distance of a bowshot." This account by Johannes Tzetzes in the 12th century is widely disputed, although Proclus in the same century is said to have repeated this feat by using a large array of mirrors to burn the fleet of Vitellius at the siege of Constantinople. Since that time many others have experimented with different types of applications for using solar energy; these include many of the greatest names in the history of science.

Focusing collectors are usually one of two types:

1. Spherical
2. Linear

Spherical shapes can develop the highest temperatures because they focus the sun's rays upon a point rather than a line as does the linear (trough-type) collector. Linear collectors are much easier to construct and can be made as long as needed. Both require a close alignment to the sun or the reflected radiation will miss the target area.

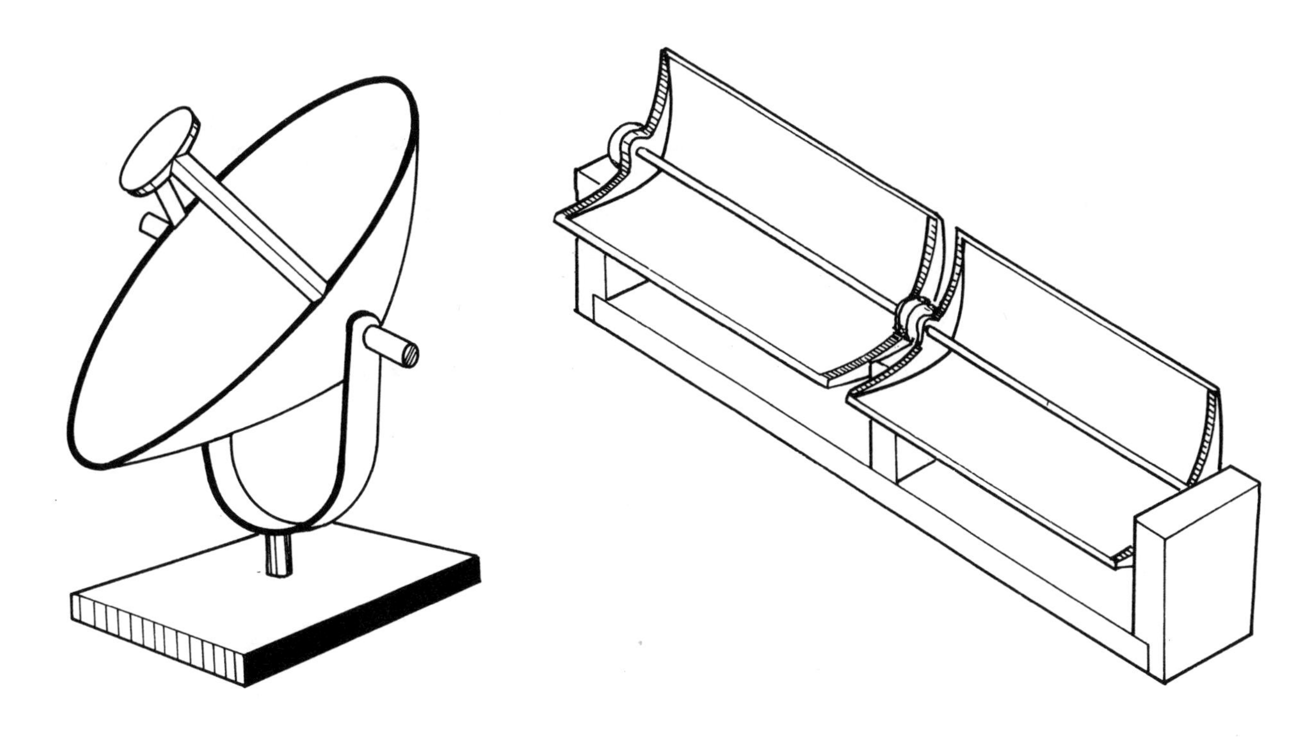

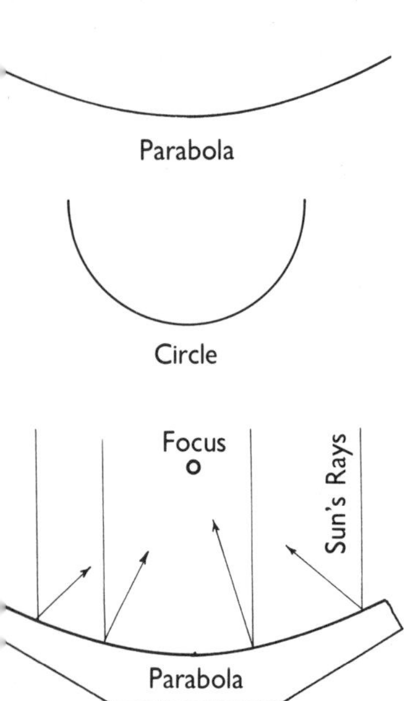

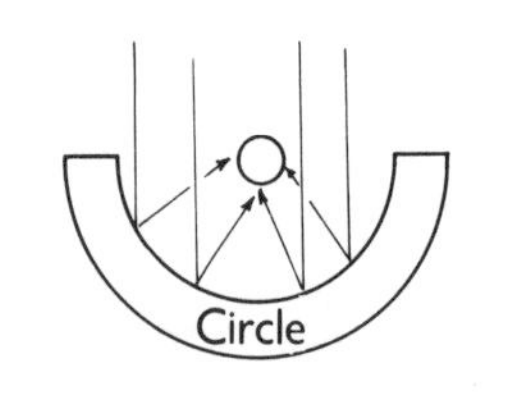

Spherical collectors consist of curved reflectors shaped to concentrate sunlight onto a small target. Curves may consist of either parabolic or circular shapes.

Looking at this illustration, it can be seen that the larger reflective area concentrates its sunlight onto the target, magnifying the intensity many times.

Hemispheres (half circles) are sometimes used as focusing collectors even though a circular shape will not focus on one point as a parabola will. This is compensated for by making the target larger.

Linear collectors often are more practical for a wide variety of uses if the target is long and thin since they can be made to adapt to any given length. If the target is sufficiently large and the parabolic or circular shape aligned, it is not always necessary to have this type of collector track the sun, saving considerable expense. Linear collectors are particularly useful for generating steam to operate turbines or for other applications not requiring temperatures over 600° F.

Focusing collectors are designed around the physical law of reflection: THE ANGLE OF INCIDENCE IS EQUAL TO THE ANGLE OF REFLECTION. This can easily be seen in the illustration.

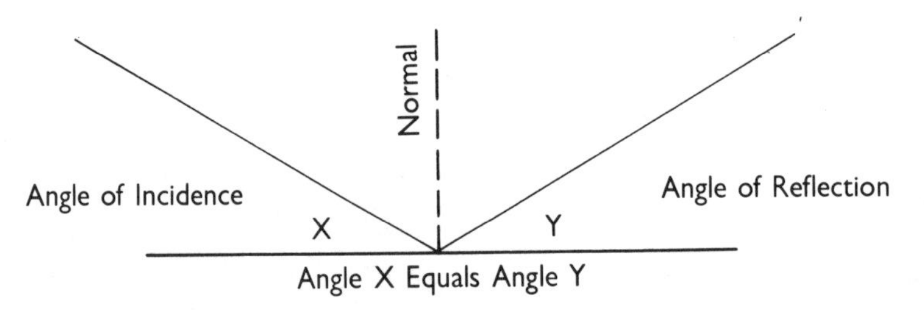

Angle X Equals Angle Y

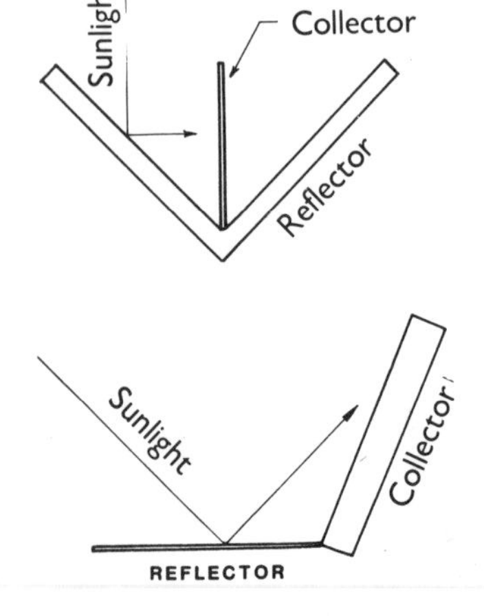

Even though the same principle applies, the reflector shapes may vary considerably from one to another.

Many factors affect the ability of a concentrating collector to reach high temperatures. These include the quality of the reflective surface, the amount of solar radiation available, the accuracy of the reflector shape or curve; the accuracy of the tracking mechanism, and the ability of the target to absorb heat. The amount of solar energy hitting the earth's surface varies considerably with both the seasons and the weather. The highest temperatures can be reached only by selecting the times and locations where the greatest amount of sunlight can be expected. Although flat-plate collectors can produce heat by indirect solar radiation, focusing collectors cannot; they must focus on the sun at all times.

Concentrating collectors can be used for a wide variety of purposes, including both industrial and scientific applications. These uses include solar furnaces, steam turbines, hot water heaters, solar cookers, solar ovens, and solar power plants.

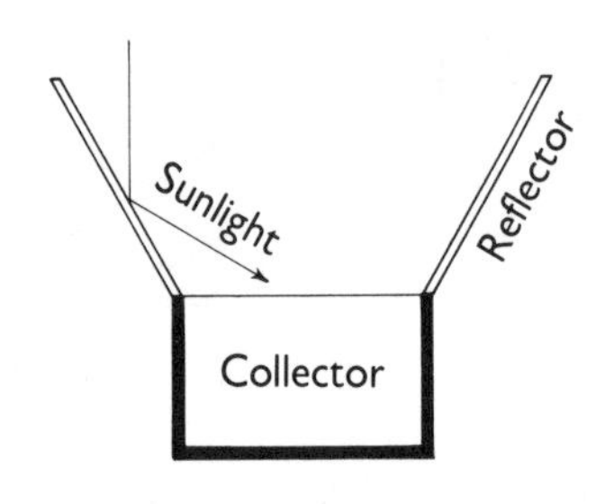

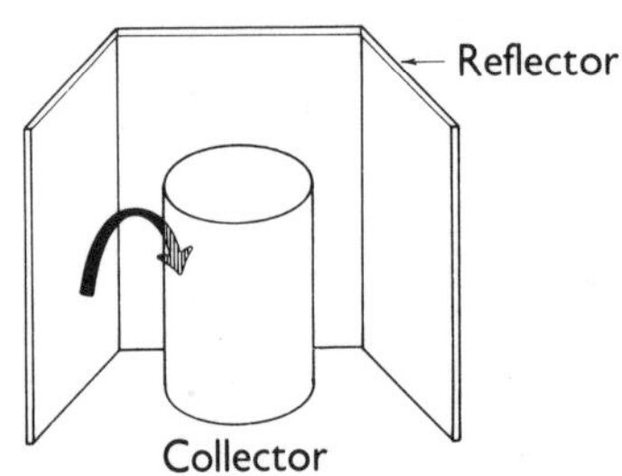

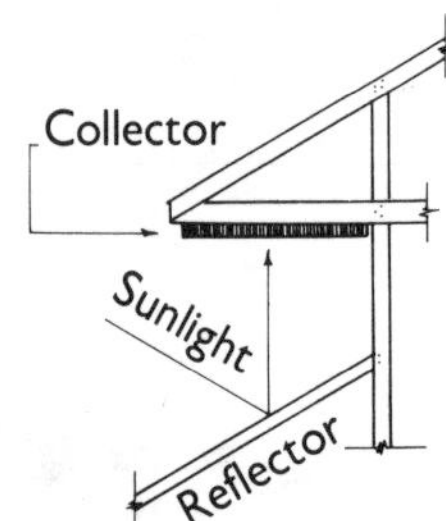

Developing Parabolic Curves

If you were able to throw a baseball straight (horizontally) as hard as you could and, at the same time, drop a baseball from the same height straight down, they would both reach the ground at the same time. However, the thrown baseball would fall in a curved path called a parabola, while the dropped baseball would fall in a straight line. If this parabolic curve were frozen in midair and precisely oriented toward the sun, it would possess a unique property: the ability to focus the parallel sun's rays onto a single point, thereby creating intense heat by using only solar radiation.

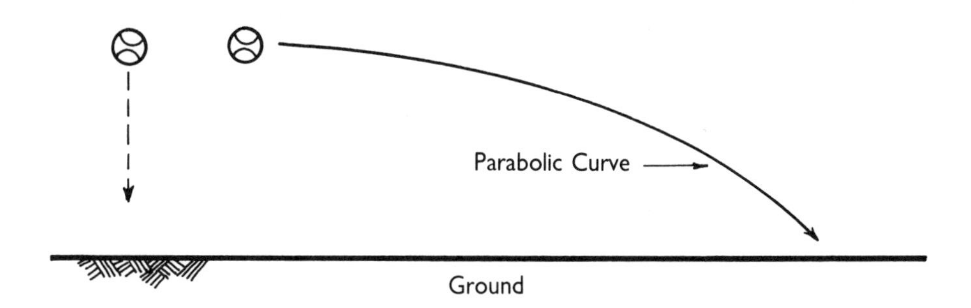

As studied by the ancient Greeks, a parabola can also be described in terms of conic sections: curves produced by planes intersecting or cutting through a right circular cone. Four types of curves may be produced within a cone.

Various methods of mathematically computing and graphically constructing parabolas exist; here are some commonly used methods.

METHOD A

A parabola is generated by a point moving so that its distances from a fixed point, referred to as the focus, and from a fixed line, known as the directrix, remain equal. In the following example, the directrix (straight line) and

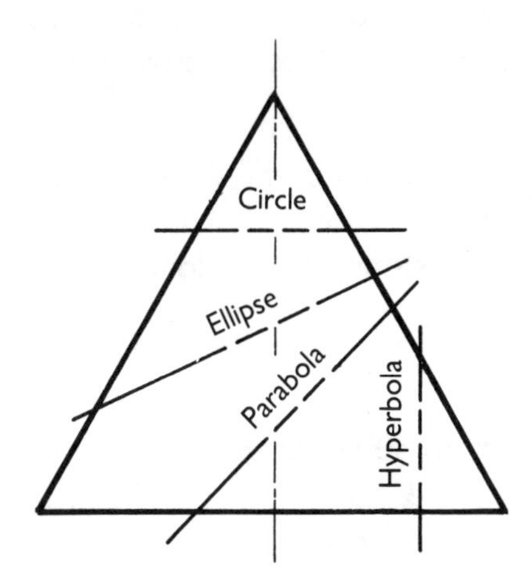

Right Circular Cone

focus (single point) have both been established, enabling the parabola to be constructed graphically.

A = A′
B = B′
C = C′
D = D′
E = E′

This is a particularly good construction method if the focus is needed at a certain point.

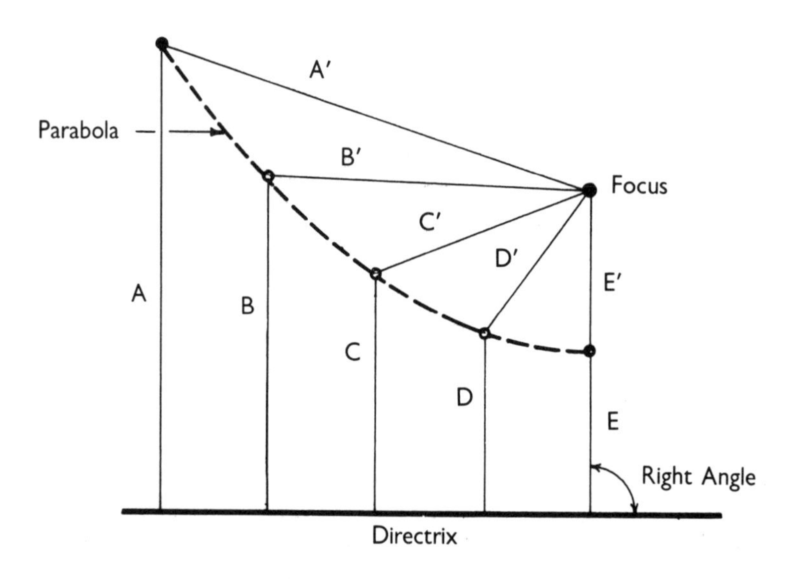

METHOD B

Another method of constructing a parabola is by using pins and strings and a T-square to develop the curve:

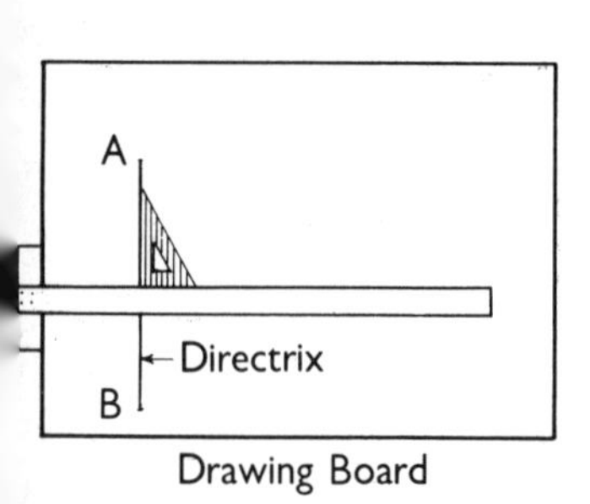

Drawing Board

1. Draw a vertical line A–B. This will be the directrix.
2. Bisect line A–B to find the center, point C.
3. Choose a convenient focal point or focus, point F, along a line drawn perpendicular to point C. Locate focal point (F) at whatever distance desired.

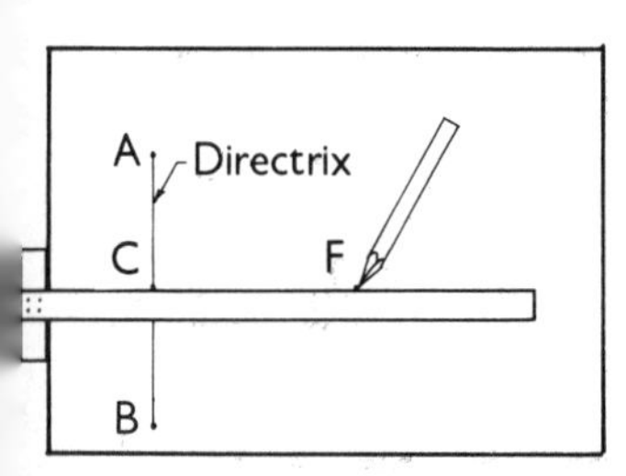

4. Choose a point L near the end of the T-square. Line the T-square up along C–F. Cut a string long enough to stretch from C to L. The farther out L is, the longer the string, and the longer the parabola will be.
5. Now attach the end of the string at point F (the focus) and the other end to the T-square at point L.
6. Put the pencil against the string along the T-square at P and trace the parabola as you slide the T-square down. Remember that line P–L must always be perpendicular to the directrix, so keep the string tight against the T-square with the pencil as shown. This gives you half of the parabola.

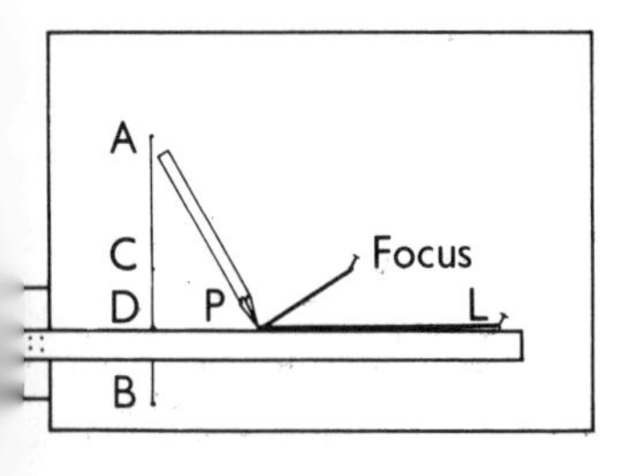

Parabolic Solar Furnace

The history of parabolic furnaces dates back to ancient Greece. Although Archimedes may have performed the first major practical use of concentrated energy by burning ships in the Roman fleet, Dositheius, a Greek mathematician in the third century B.C., discovered the basic principle of using parabolic curves as devices to focus the sun's rays on a single point and built the first parabolic mirror. Later, in the 16th century, Leonardo da Vinci became intrigued with parabolic curves and solar mirrors. He began building a giant mirror around 1515 that was estimated to be about four miles across. He never finished it, unfortunately; it was to be used for boiling water in a dyeing factory.

The solar furnace in this book was designed as a hot dog cooker to demonstrate the energy available from the sun when concentrated on a target. Solar energy produces clean heat, much cleaner than if a coal or gas furnace were used; it is an excellent source for cooking. Fairly high temperatures are possible with this furnace if the heat loss of the target object is cut down to a minimum and the focus is perfectly aligned with the sun. This can be done by shielding the solar furnace from the wind and by insulating the object on the back side (toward the sun), exposing only the areas where reflected sunlight hits.

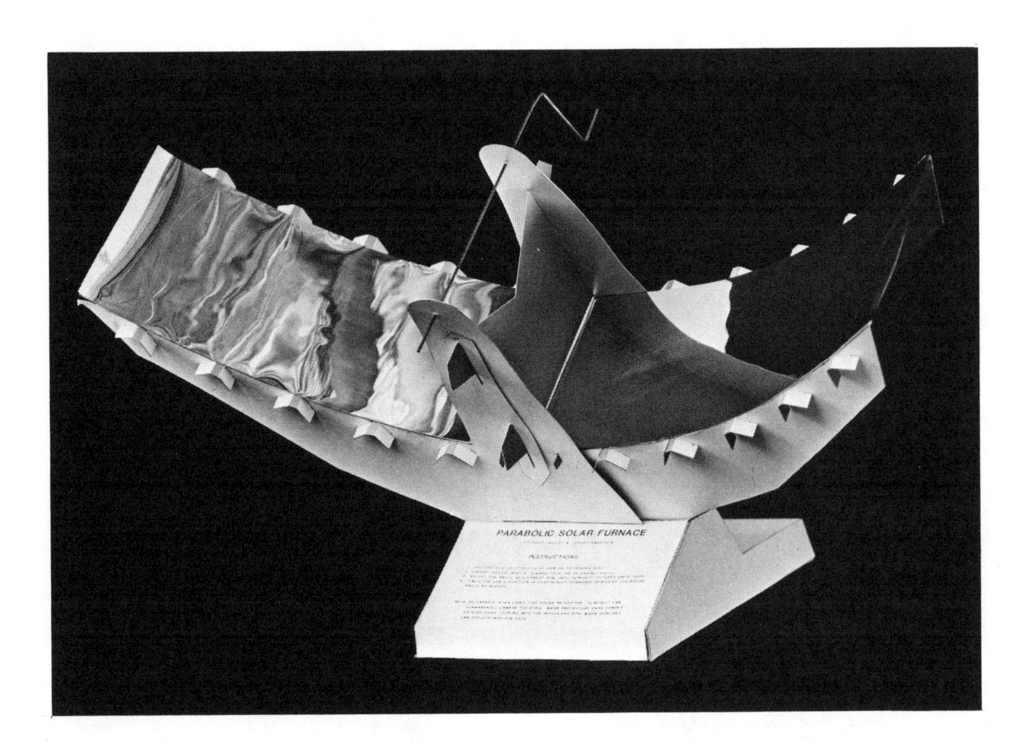

Assembly Instructions

1. Cut out pieces A, B, G, P, U. V, T and Q on solid lines.

2. Score ridge folds.

3. Score valley folds.

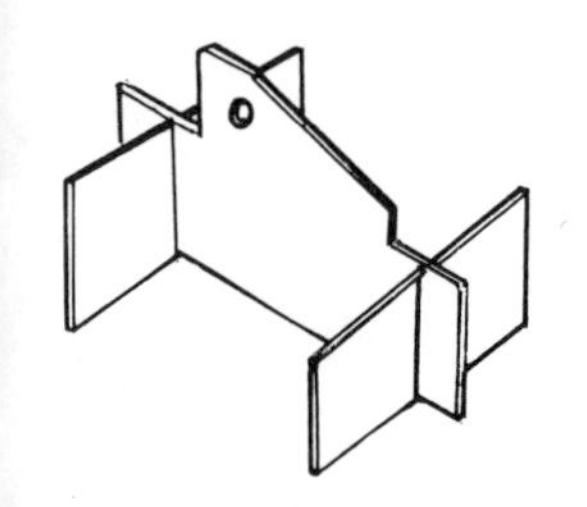

4. Cut out the center notch on the upper base assembly (piece Q) and the notches on U, V and T.

5. Punch holes at the (+) marks located on pieces P, Q and T.

6. Insert U and V into T then insert this assembled angle adjustment support into the notch of piece Q.

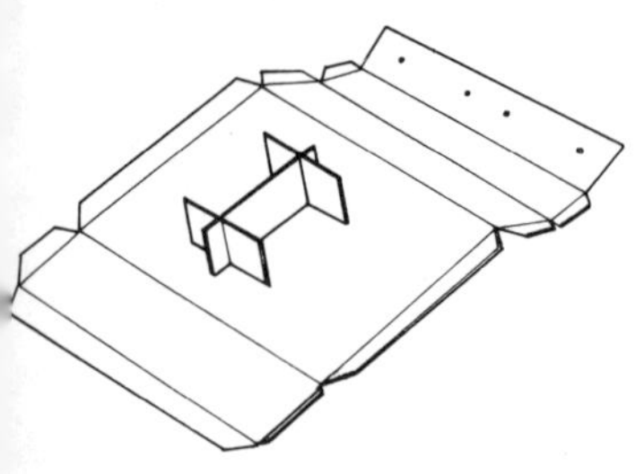

7. Assemble the base by joining tabs with glue. However, leave the "INSTRUCTIONS" portion of the upper base assembly (piece P) aside to enable access to the four aligned holes; this is where the parabola attaches to the base.

8. Cut out, score, fold and notch (where necessary) all remaining pieces. Glue piece N to a strip of cardboard before cutting it out.

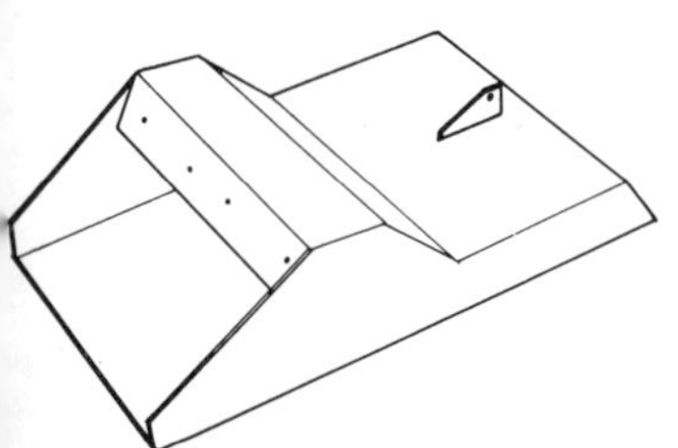

9. Assemble each side of the parabola in the following manner:

   A. Place parabola arm D beside parabola arm C and lock the five arm separators (pieces H) into the notches located on each piece.

   B. Repeat the above procedure for the other side, using parabola arms E and F. The two assemblies should be mirror images of each other.

   C. Be sure the two tabs with (+) marks located on pieces D and F are aligned on the same side; punch holes in each (+) mark.

   D. Glue the two parabolic arm dividers (pieces V and K) in place.

   E. Glue the two transverse stiffeners (pieces AA and BB) into place between the parabola arms.

   F. Glue together (back to back) the two parabolic arm dividers (pieces J and K); punch holes at the (+) marks on each piece. Insert the tabs into the remaining notches on the parabola arms. This locks the two arms together.

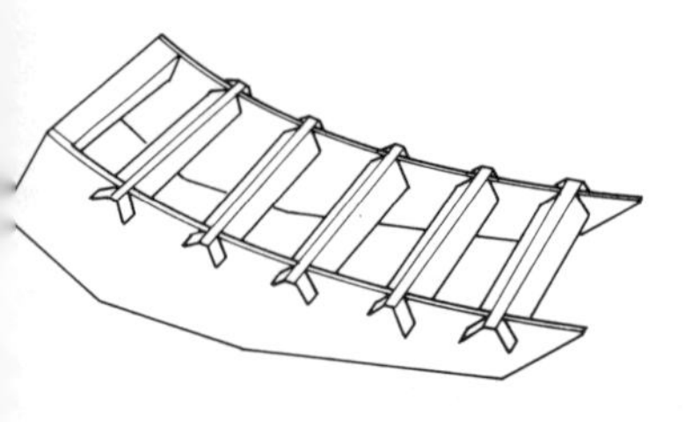

10. Attach the reflector to the base by sandwiching the two protruding tabs of D and F between the flaps on piece Q and the "INSTRUCTIONS" portion of the upper base assembly (piece P).

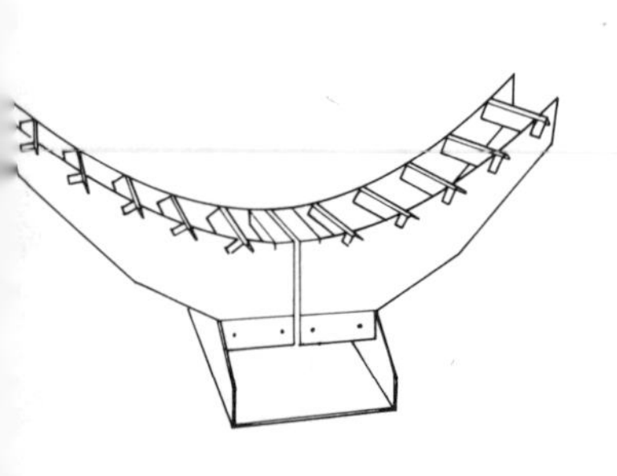

11. Insert four large paper fasteners through the aligned holes than finish gluing the upper base assembly.

12. Attach the two focus arms to the assembled parabola; the two smaller notches of each piece should fit over the tabs protruding on each side. Glue the two focus arm stiffeners (pieces Y and Z) to the focus arms (pieces R and S). Be sure all the slots are clear.

13. Insert each triangle-shaped end of the focus arm gusset (piece O) into the remaining two notches of the focus arms, then slide the two notches of

each focus arm gusset lock (piece X and piece W) over the protruding triangular ends. Slide the focus arm gusset lock down; this locks the focus arm gusset securely in place.

14. Using spray adhesive, cover the two parabolic curve pieces (M and L) with aluminized mylar, cut to fit (6" by 16"). Use two people, if possible, for this; the reflective mylar must be spread evenly to avoid wrinkles.
15. Glue each parabolic curve piece to the parabolic arms. Be sure they rest evenly (and flat) against all supports or the parabolic curve will not be accurate.
16. Punch a small pinhole in each focus arm at the point marked (+) at the top; insert a straightened coat hanger through to the other side. This is the focal point for the parabola. Clean the coat hanger with steel wool to remove all paint.
17. Punch holes through all the (+) marks of the angle adjustment arm (piece N) then attach one end to the underside of the parabola (at the punched hole) with a paper fastener; the other end attaches to the base angle adjustment support (piece T).

This completes the assembly instructions. To operate the completed parabolic reflector, follow the instructions printed on the base.

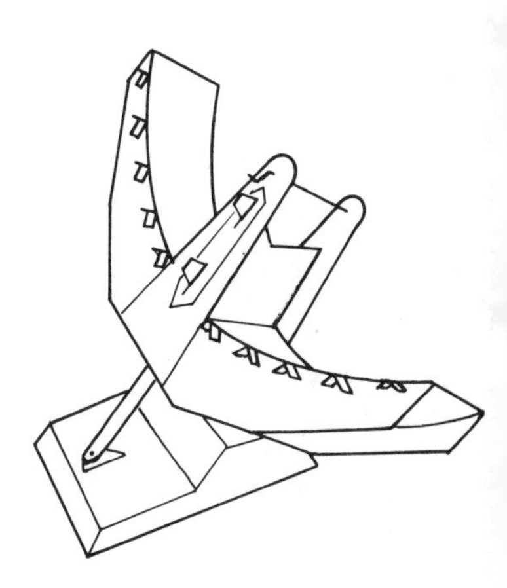

The solar furnace has a vertical adjustment so that the optimum position can be held while objects are being heated. To align the target toward the sun, orient the base until it faces the sun. Next, align the shadow cast by the upper focus arm gusset until it is parallel to the focus arm gusset lock. When it is parallel, the solar furnace will directly face the sun at that moment. Readjust it when the shadow moves, or as necessary.

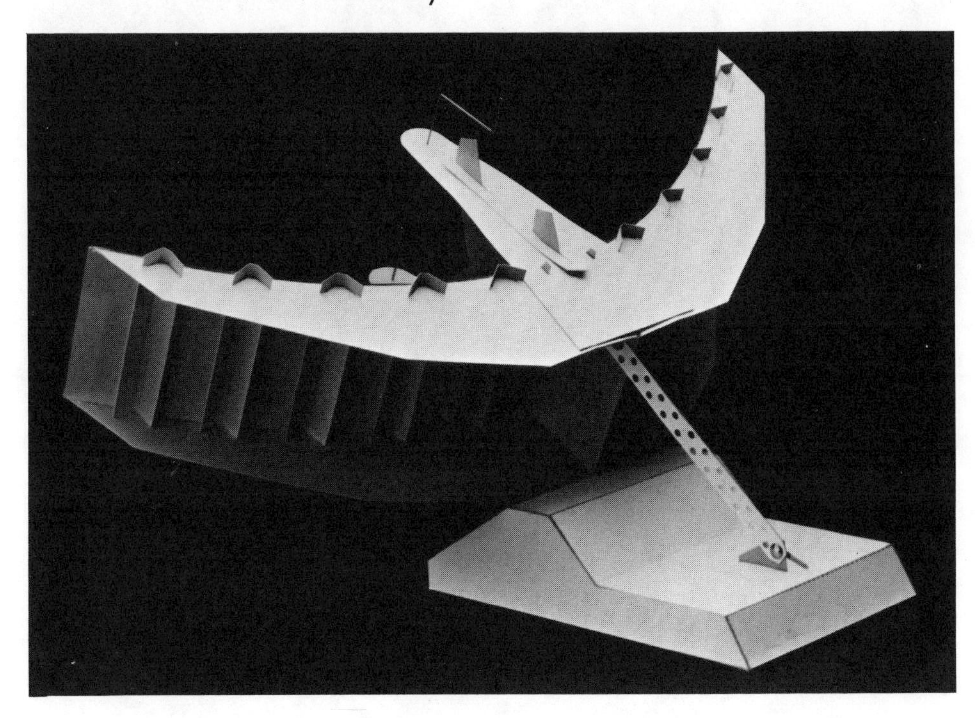

"Sunflower" Portable Parabolic Reflector

Since the surface of a circular parabola, like a sphere, is a double-curved surface, it cannot be developed from a flat surface. The "Sunflower" reflector is only an approximate parabola; it closely approximates the curved surface and has a focal point with a slightly larger target than a true parabola would ordinarily have. Nevertheless, sufficiently high temperatures can be reached to produce a strong warming effect and to allow the reader to perform some simple heat experiments when objects are placed at the focal point. For example, water can be heated if a blackened soup can is filled with water and held in place at the focal point until the maximum temperature is reached. If the top of the can is covered with an insulating material to minimize heat loss, the water will heat up even faster (and hotter). This project is smaller than the original, so an enlarged version will do much better. Portable cookers such as this have been developed for use in areas where insufficient resources are available for other types of cooking.

Since this project was designed to be portable, no provision has been made to include a support for the target area. However, different devices can be used to hold objects in place at the focal point. One simple method is to scoop out a hole in sand or dirt the approximate diameter of the reflector base then place the base in it; the reflector can then be rotated toward the sun whenever necessary. A target can be made from two forked sticks stuck in the ground on either side of the reflector with another stick suspended between them at the height of the focal point (see sketch). The focus is 7 1/16" from the base. Be extremely careful when using this reflector not to look directly into the reflected sunlight, since extensive eye damage could occur.

Assembly Instructions

1. Cut out all eight pieces (A–H) on solid lines.

2. Score optional holding tabs on the ridge fold line.

3. Punch all holes (both "•" and "+") with a paper punch; be careful to punch the holes in the exact spot or the curve will not be accurate.

4. Cover all eight pieces with highly reflective aluminized mylar or ordinary household aluminum foil. Attach the reflective surface by spraying the underside of it with a spray adhesive such as "Photomount". Smooth out the wrinkles very carefully with a soft rag, working toward the outer edge from the middle.

5. Starting with pieces A and B, and working in a counter-clockwise direction, hold the two edges together, then insert paper fasteners through the punched holes on the edge. The round head of the fastener should be placed on the inside next to the reflective surface. Continue until the next-to-last piece is reached (piece G); note that it has no holding tabs on the edge. This allows the last piece (H) to be placed on the outside.

6. After attaching piece G, place piece H on the outside of pieces A and G, then insert paper fasteners into the remaining holes. This should lock the entire reflector into an approximate parabolic curve, completing the assembly.

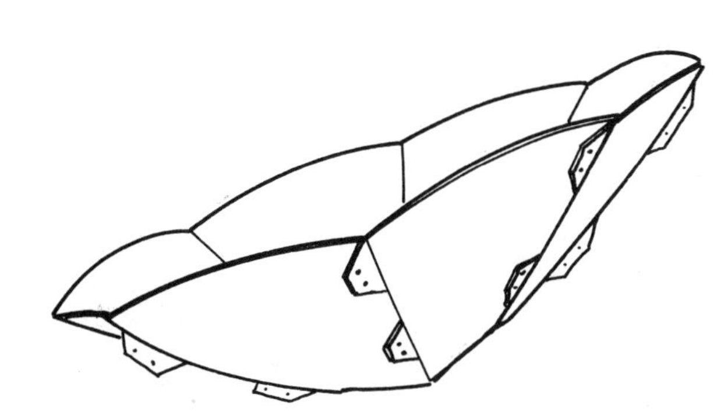

Glossary

absorber plate: that part of the solar collector which receives the incident solar radiation and transforms it into useful energy (heat). Although it is usually a solid surface used to transfer energy to the transfer fluid, it could be the transfer fluid itself (such as a "black liquid"). The absorber plate surface should be painted matte black for best results.

British thermal unit (Btu): the heat needed to raise 1 pound of water one degree Fahrenheit; this is approximately equivalent to the heat given off by burning one kitchen match. One Btu = 252 calories

calorie: the quantity of heat needed to raise the temperature of 1 gram of water one degree Centigrade

concentrating collector: a solar collector that concentrates the sun's energy through the use of lenses, reflectors or other devices onto a heat exchanger with a surface area smaller than the aperture

colatitude: latitude subtracted from 90°. For example, colatitude for Philadelphia (latitude 40°) is 50°

condensation: the process by which a vapor is changed to a liquid

conduction: the process by which heat is transferred through a solid material

convection: 1. the transfer of heat between a stationary surface and a moving fluid (liquid or gas). 2. the transfer of heat within a fluid created by movements (convection currents) within the fluid

density: weight of a substance in pounds per cubic foot of volume

dial plate: divides the time of day into equal parts, or hours, on a sundial. Time is determined by observing where the shadow cast by the gnomon falls onto the dial plate

energy: the capacity for doing work

equation of time: the time difference between local apparent time and local mean time.

greenhouse effect: process by which glass allows short wavelength solar radiation to enter but prevents long wavelength radiation (heat) from leaving

gnomon: the object which casts a shadow of the sun onto the timepiece or dial plate

heat: the energy an object generates due to the motion of its molecules

heat loss: a reduction in thermal energy occurring when a difference in temperature is created through conduction, convection, or radiation

incident angle: the angle between the sun's rays and the surface of the flat plate solar collector.

insolation: the rate of solar radiation received by a unit surface area in unit time. (Btu/h.ft.^{2} or W/M^2) h = outside surface heat transfer coefficient. W = watts. M = meters

joule: a unit of energy or work which is equivalent to one watt second; it is one newton acting for a distance of 1 meter and is equal to 0.737 foot pounds

langley: a measure of solar insolation or radiation. One langley equals one calorie of radiant energy per square centimeter. One langley per minute equals 221 Btu per square foot per hour

latitude: the location of a specific point with reference to its distance north or south of the equator.

Local Apparent Time (LAT): the time shown on most sundials; time is determined by the sun's angle

Local Mean Time (LMT): the time a sundial would show if the sun moved across the sky at a constant rate from season to season

longitude: the angular distance east or west of Greenwich, England on the earth's surface; longitudinal lines pass through the celestial (north-south) poles

meridian: an imaginary north-south line on the earth's surface

parabola: a curve or conic section produced by the intersection of a cone and a plane parallel to the curve's side

radiation: the transfer of heat which a body gives off in the form of electromagnetic waves

solar collector: a device used to obtain by absorption incident solar radiation and which transfers the energy to a fluid (liquid or gas) that flows to a storage system. It has three main components: a transparent cover (glass or plastic), an absorber plate (absorbent surface), and the heat transfer medium (usually water or air)

standard time: the local mean time of a single meridian which is adopted by a zone of territory on either side of the meridian so that the same time can be used over a wide area

temperature: the amount of heat in anything, usually measured in terms of hotness or coldness from a standard scale

transfer fluid: the liquid or gas that carries energy obtained by the sun away from the collector and in and out of the storage system. The most common liquids used are water or a water-ethylene-glycol solution; the most common gas is air

transparent cover: the barrier against convection and thermal radiation on a flat-plate solar collector which protects the internal components against dirt and precipitation. The most common transparent covers are 3/16" window glass and tedlar or mylar clear plastic; glass lasts much longer but is four to eight times more expensive. Polyethylene plastic film is very inexpensive but lasts for only one to two years

Trombe wall: a storage wall (usually masonry) for solar radiation used in passive solar construction. It was named after Felix Trombe, a French solar scientist and inventor

work: a process employing energy to perform some transformation or rearrangement of matter

Bibliography

Applied Solar Energy: an Introduction, by Aden Meinel and Marjorie Meinel. Addison-Wesley Publishing Company, Jacob Way, Reading, MA 01867 (1976). 651 pp. $23.95.

An invaluable aid to the layperson, as well as the serious researcher interested in the solar energy field. Intended as a textbook for seniors and graduate students in energy conservation courses, the book provides a broad perspective and a well-researched view of the uses of solar energy.

Build Your Own Solar Water Heater, by Stu Campbell and Doug Taff. Garden Way Publishing, Charlotte, Vermont 05445 (1978). 109 pp. $14.95. Paperback, $7.95.

This practical "how-to" guide details heating with a solar water heater. The homeowner will find it especially useful and easy to understand. Many diagrams clearly illustrate the construction and installation of domestic hot water solar heating systems.

Designing and Building a Solar House: Your Place in the Sun, by Donald Watson. Garden Way Publishing, Charlotte, Vermont 05445 (1977). 281 pp. $14.95. Paperback, $10.95.

This book provides factual information on both active and passive design concepts for solar houses. It includes design and construction checklists, site planning, and principles of building for maximum energy conservation. The book is comprehensive, instructive and well-illustrated with photographs and diagrams.

Direct Use of the Sun's Energy, by Farrington Daniels. Ballantine, 201 East 50th Street, New York, NY 10022 (1974). Paperback, $1.95.

A classic in the field of solar energy literature, this book presents a thorough examination of all aspects of solar energy applications currently in use worldwide. Until his death, Dr. Daniels was a leading authority on solar energy in the United States, and his book is still considered one of the best written on the subject. It is easy to read, well-researched and an important basic guide for all solar energy enthusiasts.

Energybooks 1 & 2, edited by John Prenis. Running Press, 125 South 22nd Street, Philadelphia, PA 19103 (1975, 1977). Paperback, 112 pp. $4; 128 pp. $5.

Two sourcebooks of information on alternative energy forms. Articles included cover a wide range of sources of alternative energy such as wind, tides, solar, methane and geothermal. The scope and variety of the articles gathered in the two books present a wealth of information in a way that will interest and entertain the lay person and the backyard experimenter alike.

Energy Primer, edited by Richard Merrill and Thomas Gage. Portola Institute, 485 Hamilton Avenue, Palo Alto, CA 94301 (1978). 200 pp. Paperback, $5.50.

A collection of semi-technical articles on renewable forms of energy. The primer offers access information on energy hardware, available energy literature, practical tools. The sections detailing energy forms such as solar, water, wind and biofuels are well-written and thorough.

Mechanical Drawing, by Thomas French, Carl Svensen, Jay Helsel, and Byron Urbanick. McGraw Hill Book Company, 1221 Avenue of the Americas, New York, NY 10020 (1980). 568 pp. $16.80.

This latest edition of French and Svensen's classic textbook on technical drawing is well-illustrated with colorful examples, which help clarify the graphic concepts. It can be easily adapted to high school or college level students, as well as the interested layperson. It is a comprehensive, definitive guide to mechanical drawing.

Natural Energy Workbook #2, by Peter Clark with Judy Landfield. Visual Purple, Box 996, Berkeley, CA 94701 (1976). 128 pp. Paperback, $3.95.

This is an enlightening, educational workbook on natural energy sources; it contains many simple experiments which demonstrate basic energy principles. Active participation by the reader throughout is encouraged, making the workbook thought-provoking and fun.

Other Homes and Garbage, by Jim Leckie, Gil Masters, Harry Whitehouse, and Lily Young. Sierra Club Books, 530 Bush Street, San Francisco, CA 94108 (1975). 320 pp. Paperback, $9.95.

An experimental workshop at Stanford University organized around concepts of self-sufficient living systems produced this book. It provides technical information on alternative energy systems which offer an opportunity for increased self-reliance.

Solar Dwelling Design Concepts, by the AIA Research Corporation. U.S. Government Printing Office, Division of Public Documents, Washington, D.C. 20402 (1976). 146 pp. Paperback, $2.30. Stock #023-000-00334-1.

This government publication was made available to increase public awareness and understanding of architectural solar heating and cooling concepts. It provides historic solar applications, discusses existing solar houses, and provides factual information on dwelling and site design concepts.

Solar Energy for Northwest Buildings, by John Reynolds. Center for Environmental Research, School of Architecture and Applied Arts, University of Oregon, Eugene, OR 97403 (1974). 70 pp. Spiral bound, $3.90.

This booklet presents information on solar energy and building, primarily as it relates to the northwestern United States. Various regional influences on architecture are discussed, and solar houses built in the northwest around 1967 are analyzed. This is a useful and informative guide to solar energy applications in the northwest.

The Solar Home Book, by Bruce Anderson with Michael Riordan. Brick House, 3 Main Street, Andover, MA 01810 (1976). 297 pp. $13.95. Paperback, $9.50.

This is one of the best books on solar home design; it should be required reading for anyone planning to build a solar home. Among the topics discussed are direct and indirect solar heating, soft technology approaches, and solar applications to existing houses.

Sundials, by Frank Cousins. John Baker Publishers Limited, 4, 5 & 6 Soho Square, London WIV6AD (1972). $22.50 approximately (£7.25 net).

A clear explanation of the art and science of gnomonics, with excellent photographs of historic sundials. Most kinds of sundials are discussed; excellent drawings supplement the somewhat scholarly text.

Sundials: Their Theory and Construction, by Albert Waugh. Dover Publications, Inc., 180 Varick Street, New York, NY 10014 (1973). 228 pp. Paperback, $3.50.

An excellent introduction to the theory and construction of sundials. It clearly illustrates all the common types of sundials and provides an historical and current view.

The Sun: Our Future Energy Source, by David McDaniels. John Wiley and Sons, 605 Third Avenue, New York, NY 10016 (1979, 1980). 271 pp. Text Edition, $18.95. Paperback, $14.95 (1979).

This textbook originated from Dr. McDaniels' lecture notes prepared for an introductory course on solar energy at the University of Oregon, where Dr. McDaniels is Professor of Physics. It discusses steadily dwindling energy resources and provides a clear and extensive overview of the way in which vital energy can be obtained from the sun.

Sunset Home Owner's Guide to Solar Heating, prepared by the editors of Sunset Books. Lane Publishing Company, Menlo Park, CA 94025 (1978). 96 pp. Paperback, $2.95.

This well-prepared, non-technical book explains how solar heating works for homes. It is illustrated with fine drawings and photographs which supplement the informative text. This is one of the best low-cost books on the solar market.

Technical Drawing, by Frederick Giesecke, Alva Mitchell, Henry Spencer, and Ivan Hill. Macmillan Company, 866 Third Avenue, New York, NY 10022 (1980). 880 pp. $21.95

One of the finest and most comprehensive technical drawing books available. It has detailed information on every major mechanical drawing concept. Some of the drawing procedures can be rather complex, but a conscientious student will be able to develop almost any type of technical drawing with help from this text.

SUN ANGLE VIEWER

WITH SCALES FOR HORIZONTAL AND VERTICAL SUN ANGLES

PROJECT 1

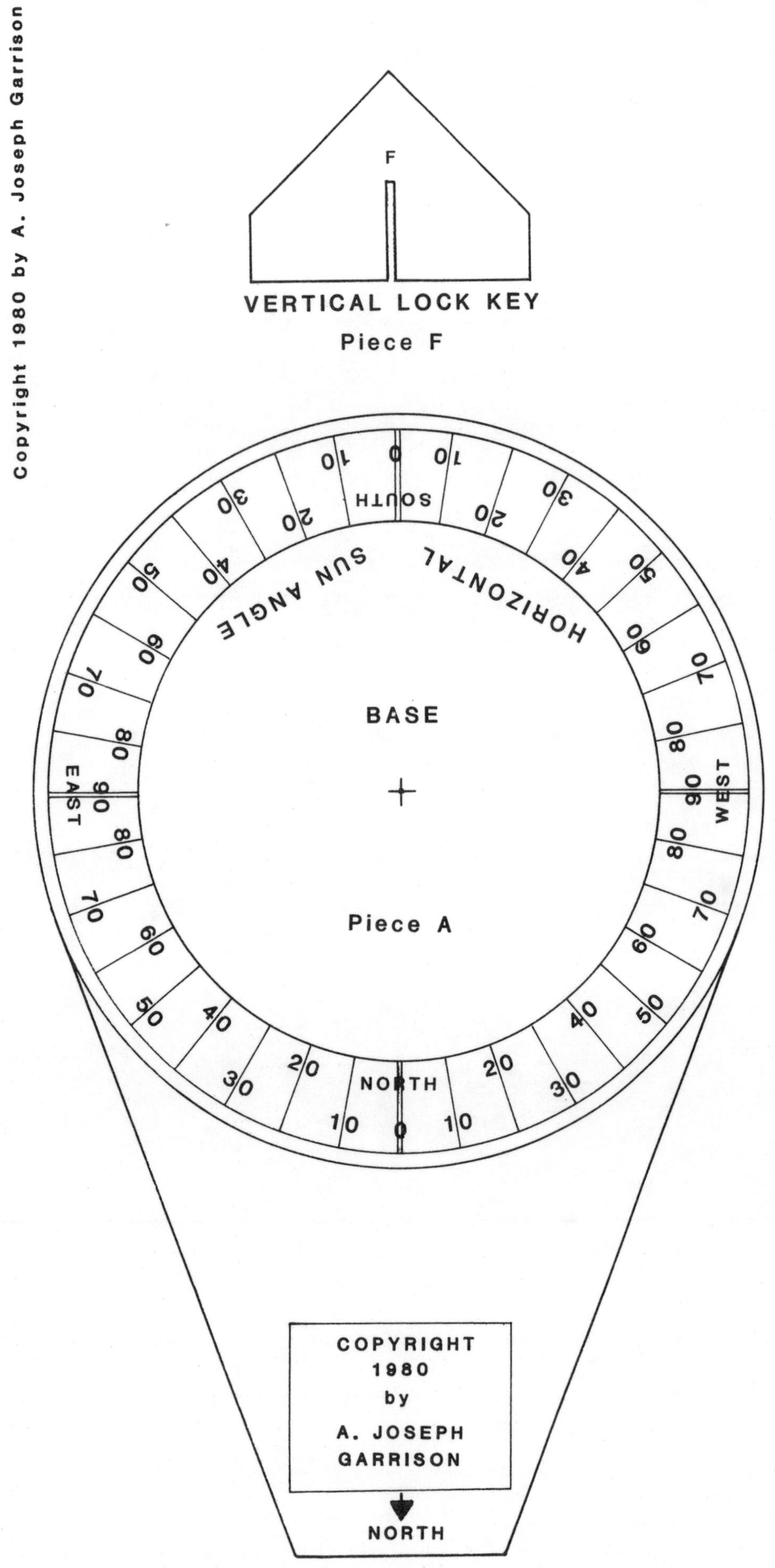

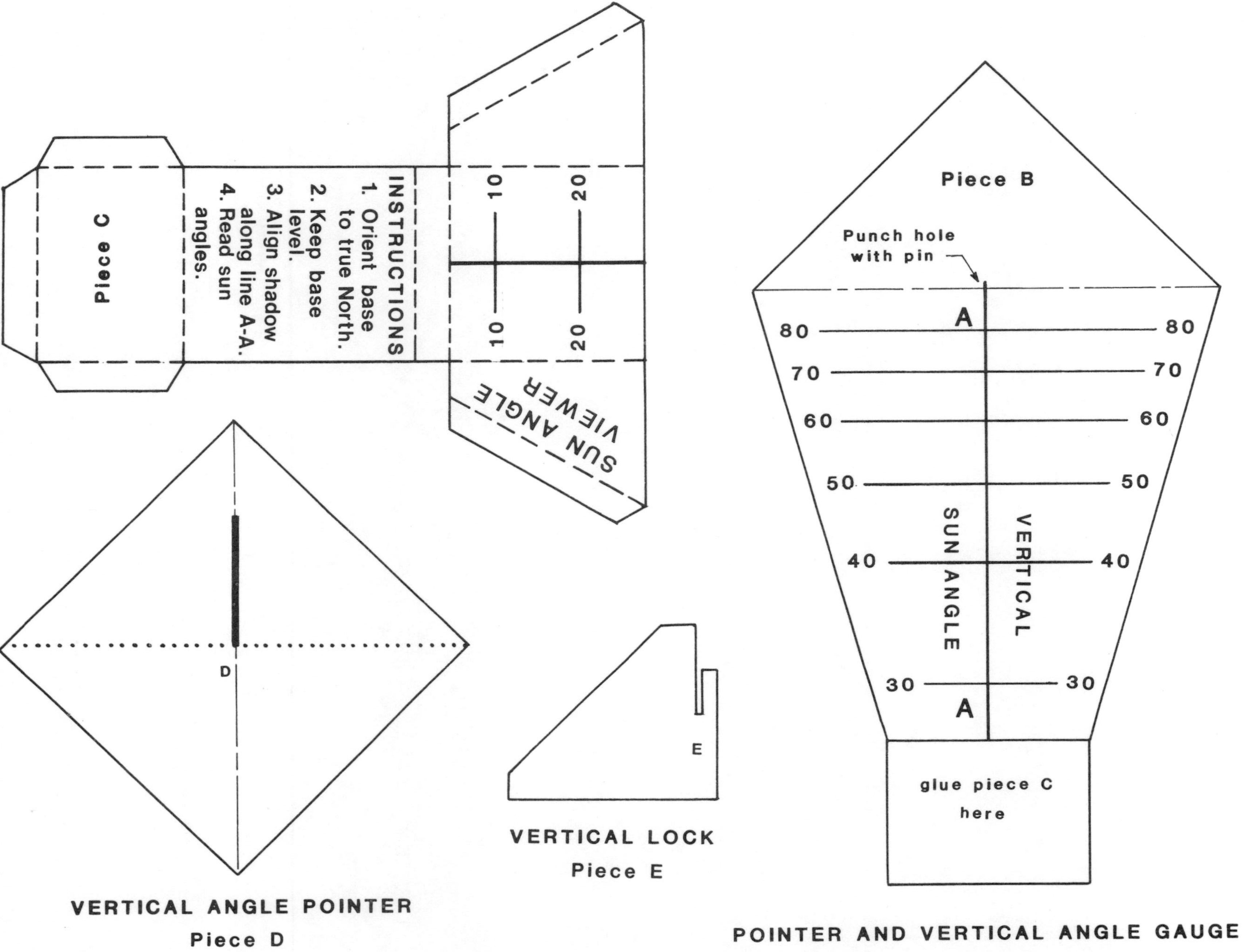

VERTICAL ANGLE POINTER
Piece D

VERTICAL LOCK
Piece E

POINTER AND VERTICAL ANGLE GAUGE

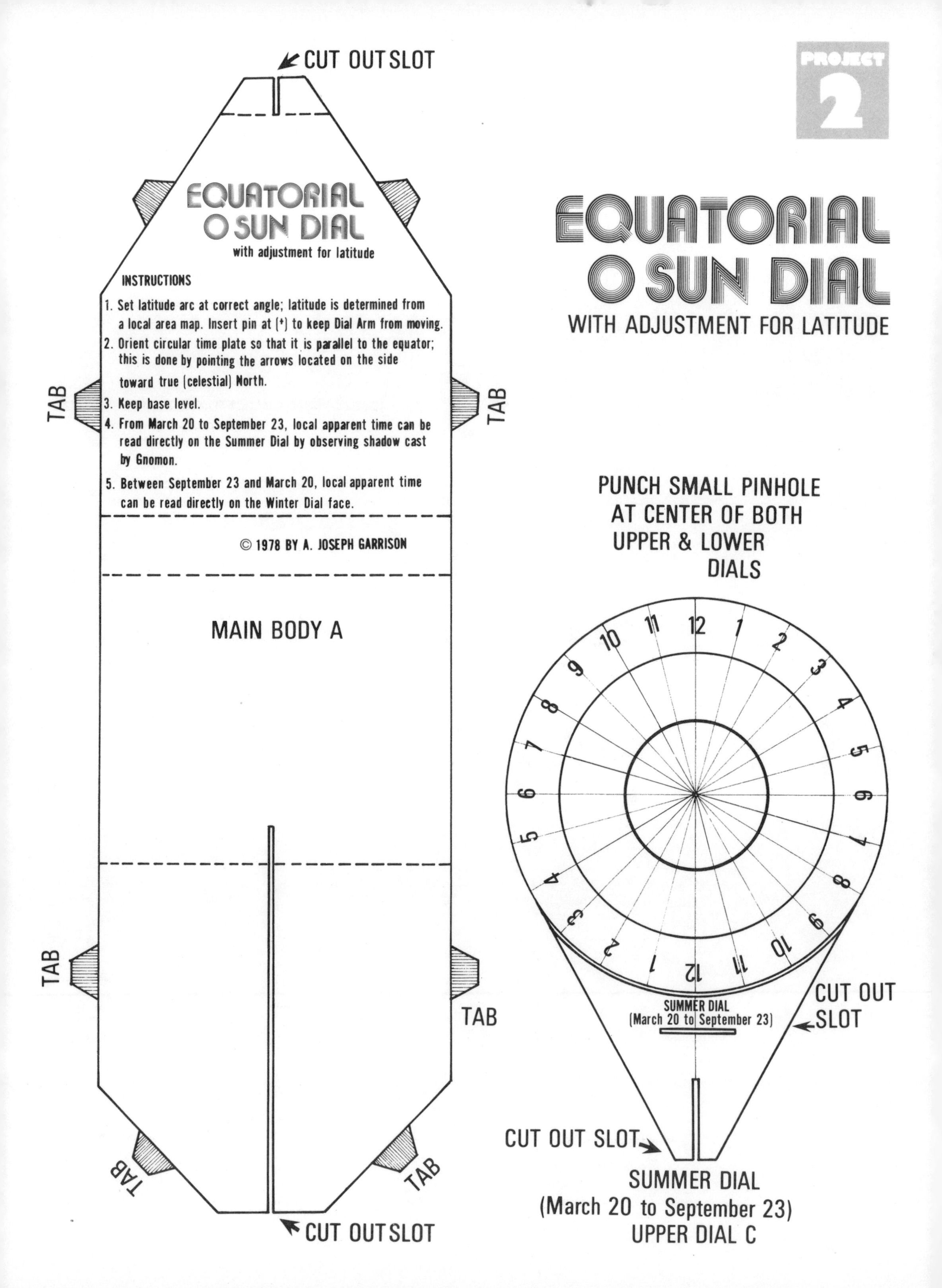

CUT OUT SLOT
PROJECT 2
EQUATORIAL
O SUN DIAL
with adjustment for latitude
INSTRUCTIONS
1. Set latitude arc at correct angle; latitude is determined from a local area map. Insert pin at (+) to keep Dial Arm from moving.
2. Orient circular time plate so that it is parallel to the equator; this is done by pointing the arrows located on the side toward true (celestial) North.
3. Keep base level.
4. From March 20 to September 23, local apparent time can be read directly on the Summer Dial by observing shadow cast by Gnomon.
5. Between September 23 and March 20, local apparent time can be read directly on the Winter Dial face.
© 1978 BY A. JOSEPH GARRISON
MAIN BODY A
TAB
TAB
TAB
TAB
TAB
TAB
CUT OUT SLOT
EQUATORIAL
O SUN DIAL
WITH ADJUSTMENT FOR LATITUDE
PUNCH SMALL PINHOLE AT CENTER OF BOTH UPPER & LOWER DIALS
12 1 2 3 4 5 6 7 8 9 10 11
12 1 2 3 4 5 6 7 8 9 10 11
SUMMER DIAL
(March 20 to September 23)
CUT OUT SLOT
CUT OUT SLOT
SUMMER DIAL
(March 20 to September 23)
UPPER DIAL C

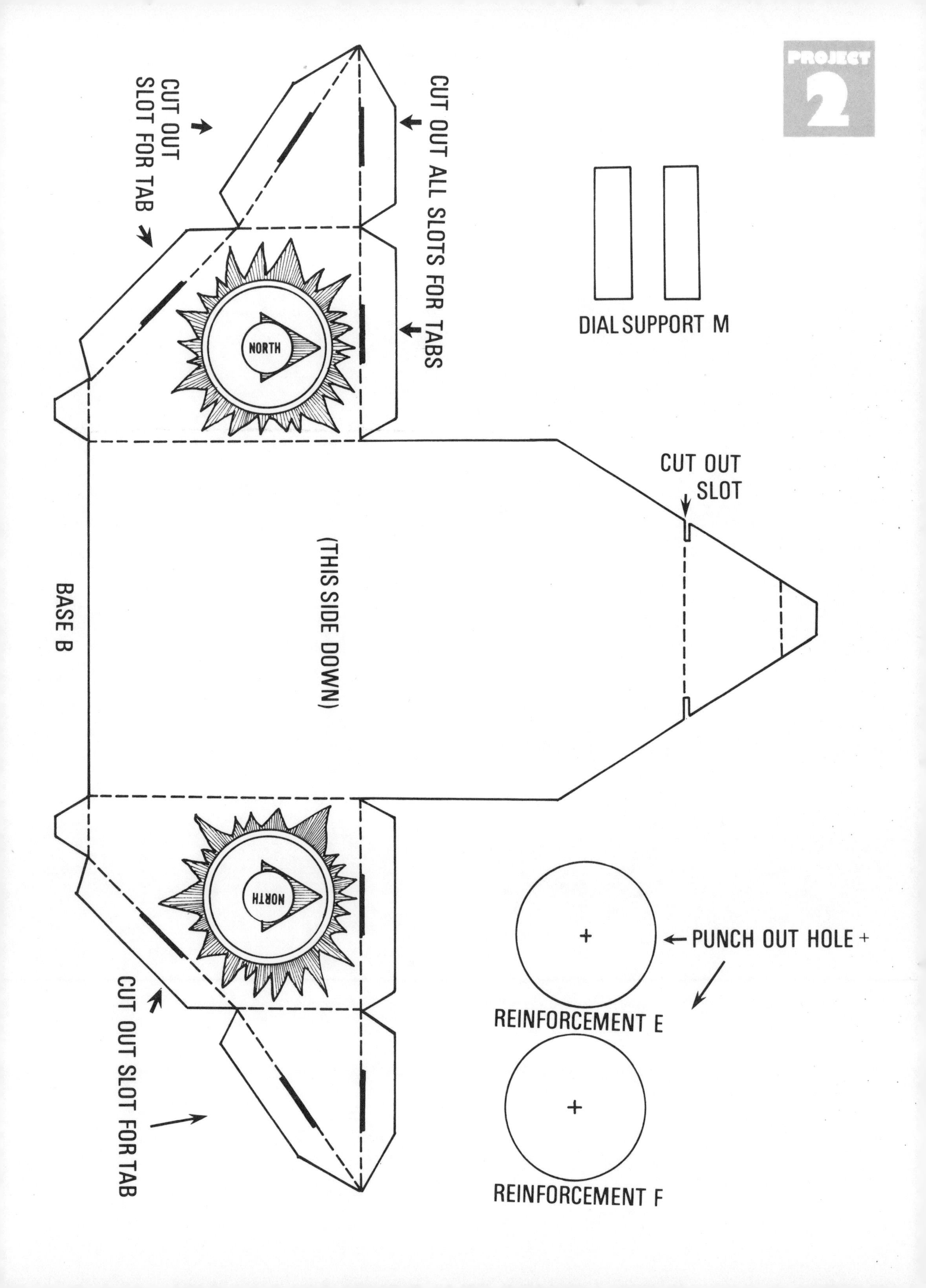
PROJECT 2
CUT OUT SLOT FOR TAB
CUT OUT ALL SLOTS FOR TABS
NORTH
DIAL SUPPORT M
CUT OUT SLOT
(THIS SIDE DOWN)
BASE B
NORTH
CUT OUT SLOT FOR TAB
PUNCH OUT HOLE +
REINFORCEMENT E
REINFORCEMENT F

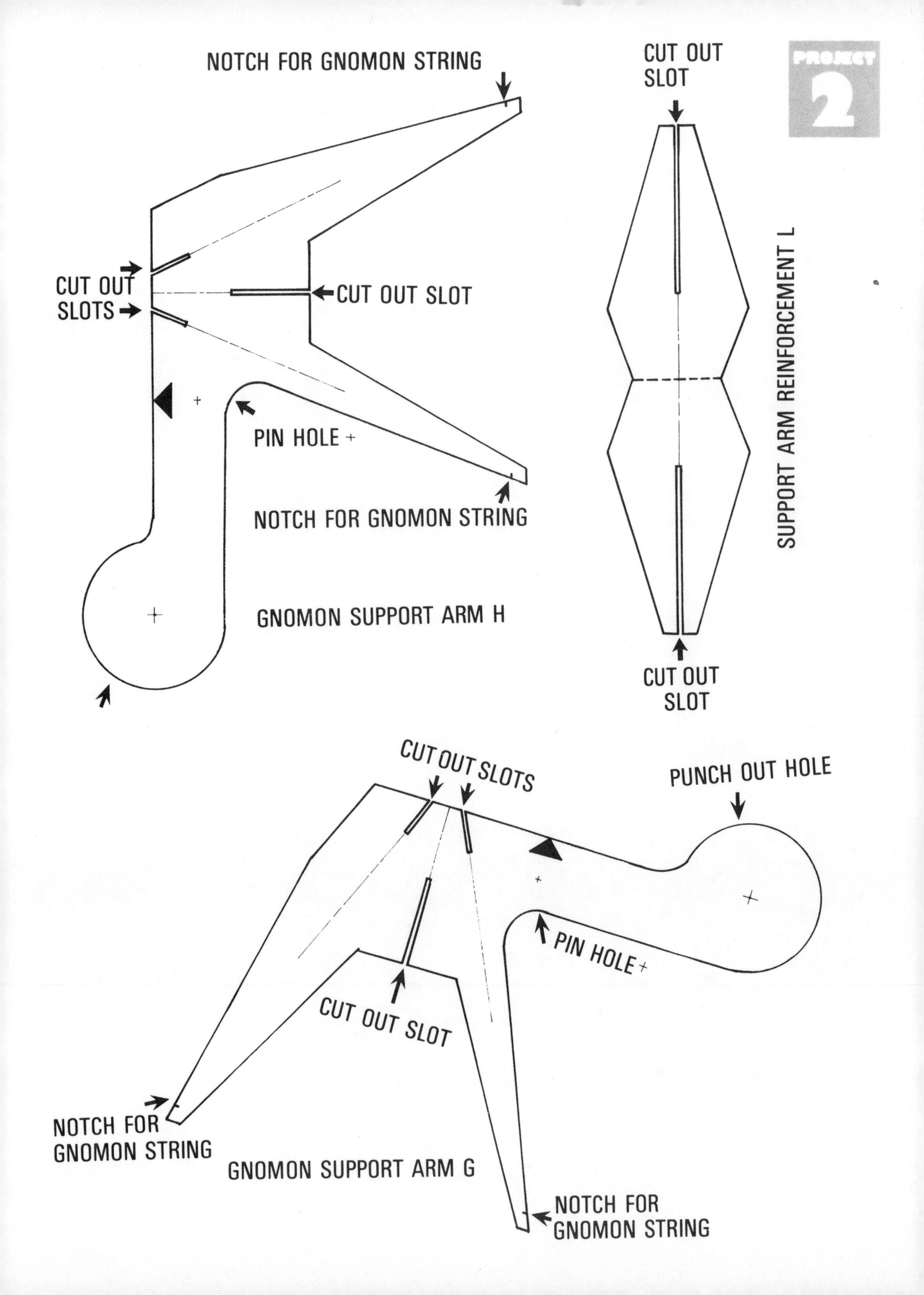
PROJECT
2
NOTCH FOR GNOMON STRING
CUT OUT SLOTS
CUT OUT SLOT
PIN HOLE +
NOTCH FOR GNOMON STRING
GNOMON SUPPORT ARM H
CUT OUT SLOT
SUPPORT ARM REINFORCEMENT L
CUT OUT SLOT
CUT OUT SLOTS
PUNCH OUT HOLE
PIN HOLE +
CUT OUT SLOT
NOTCH FOR GNOMON STRING
GNOMON SUPPORT ARM G
NOTCH FOR GNOMON STRING

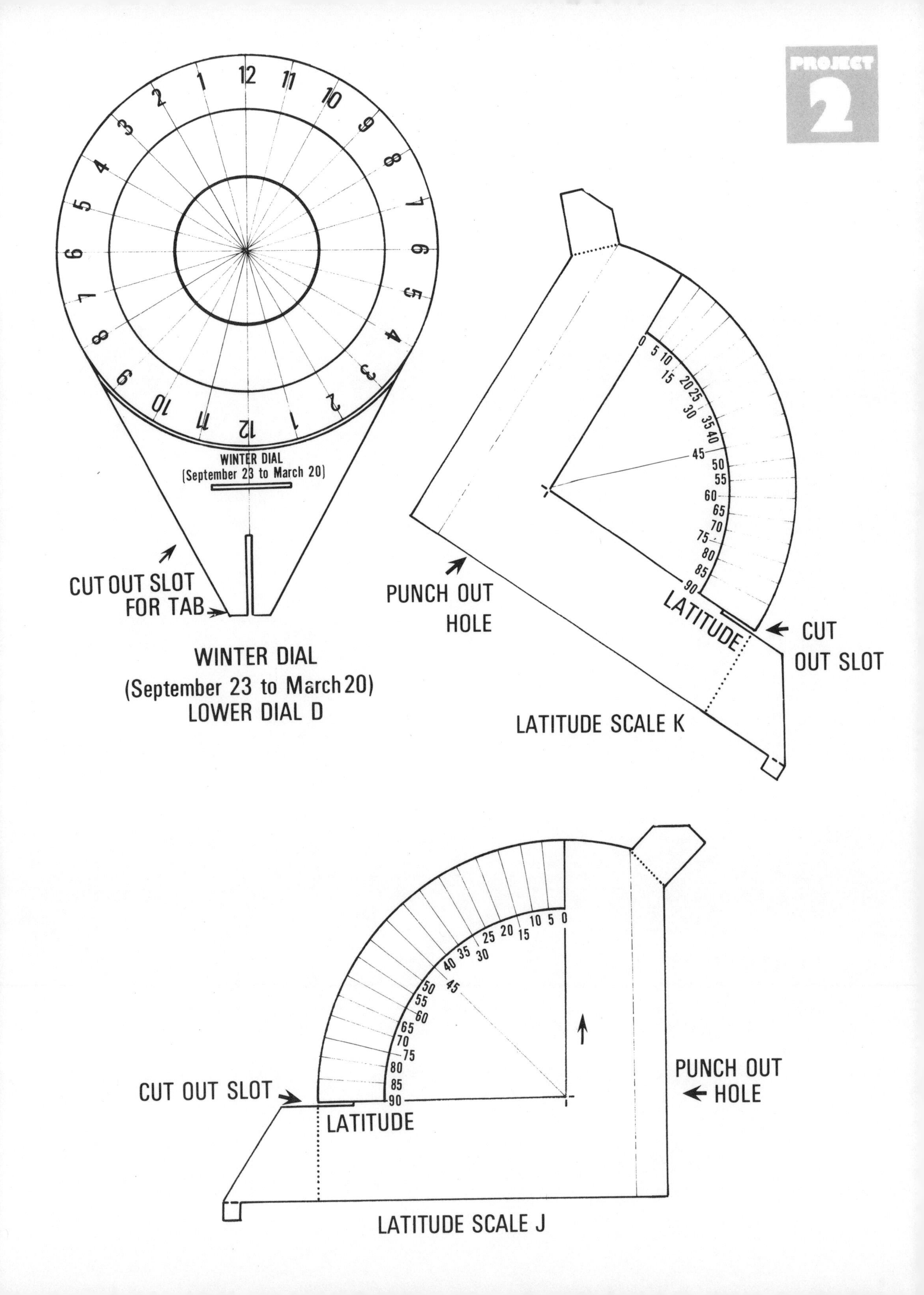
PROJECT
2
12
11
10
9
8
7
6
5
4
3
2
1
12
11
10
9
8
7
6
5
4
3
2
1
WINTER DIAL
(September 23 to March 20)
CUT OUT SLOT
FOR TAB
WINTER DIAL
(September 23 to March 20)
LOWER DIAL D
0
5
10
15
20
25
30
35
40
45
50
55
60
65
70
75
80
85
90
LATITUDE
PUNCH OUT
HOLE
CUT
OUT SLOT
LATITUDE SCALE K
0
5
10
15
20
25
30
35
40
45
50
55
60
65
70
75
80
85
90
CUT OUT SLOT
LATITUDE
PUNCH OUT
HOLE
LATITUDE SCALE J

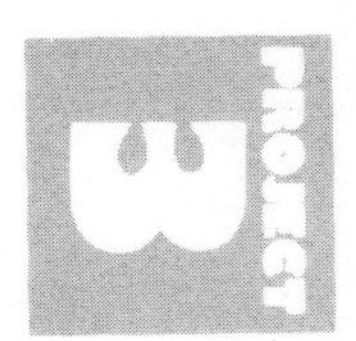
PROJECT
3

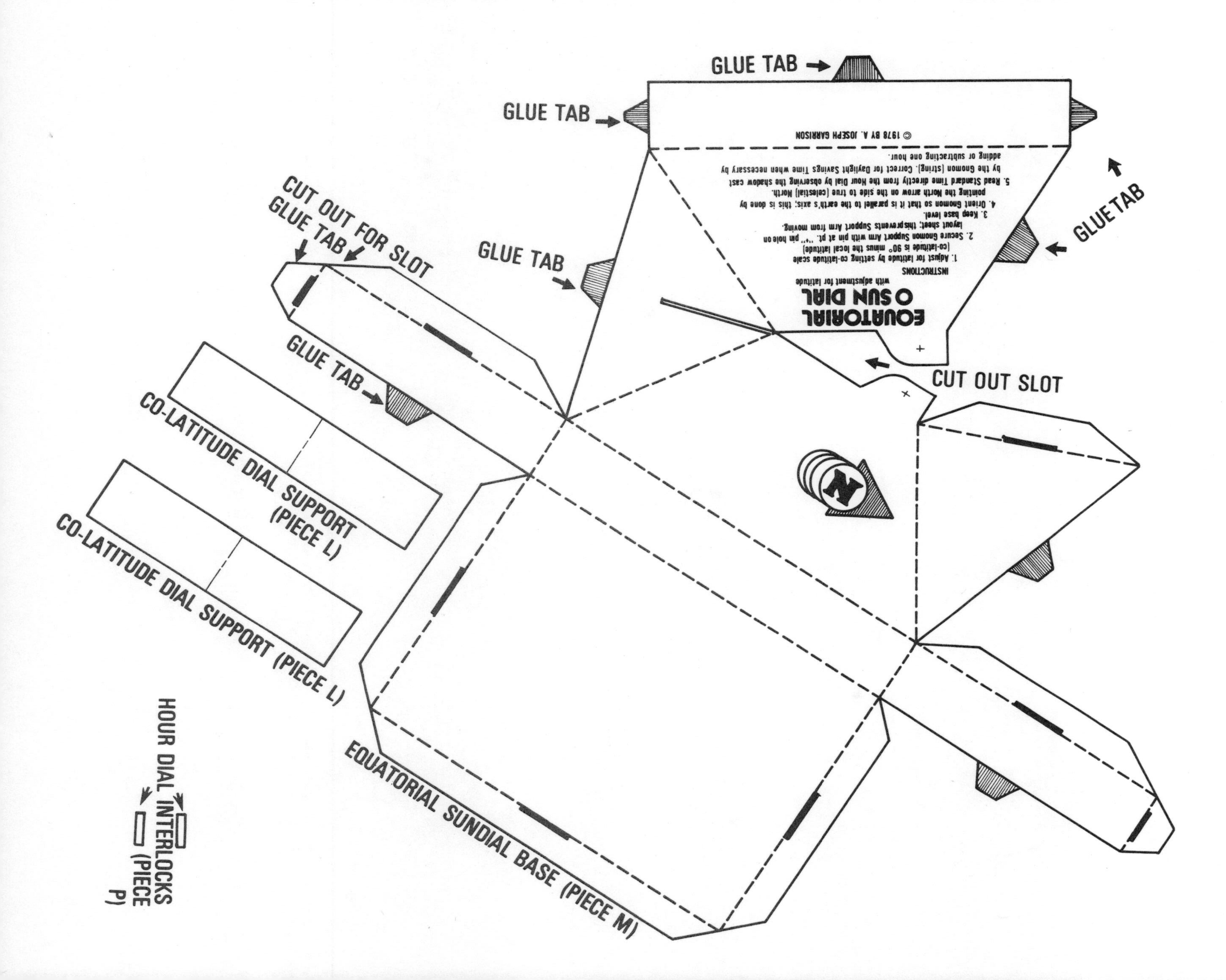
EQUATORIAL
O SUN DIAL
with adjustment for latitude
INSTRUCTIONS
1. Adjust for latitude by setting co-latitude scale (co-latitude is 90° minus the local latitude)
2. Secure Gnomon Support Arm with pin at pt. "+" pin hole on layout sheet; thisprevents Support Arm from moving.
3. Keep base level.
4. Orient Gnomon so that it is parallel to the earth's axis; this is done by pointing the North arrow on the side to true (celestial) North.
5. Read Standard Time directly from the Hour Dial by observing the shadow cast by the Gnomon (string). Correct for Daylight Savings Time when necessary by adding or subtracting one hour.
© 1978 BY A. JOSEPH GARRISON
GLUE TAB
GLUE TAB
GLUE TAB
GLUE TAB
CUT OUT SLOT
N
CUT OUT FOR SLOT
GLUE TAB
GLUE TAB
CO-LATITUDE DIAL SUPPORT (PIECE L)
CO-LATITUDE DIAL SUPPORT (PIECE L)
EQUATORIAL SUNDIAL BASE (PIECE M)
HOUR DIAL INTERLOCKS (PIECE P)

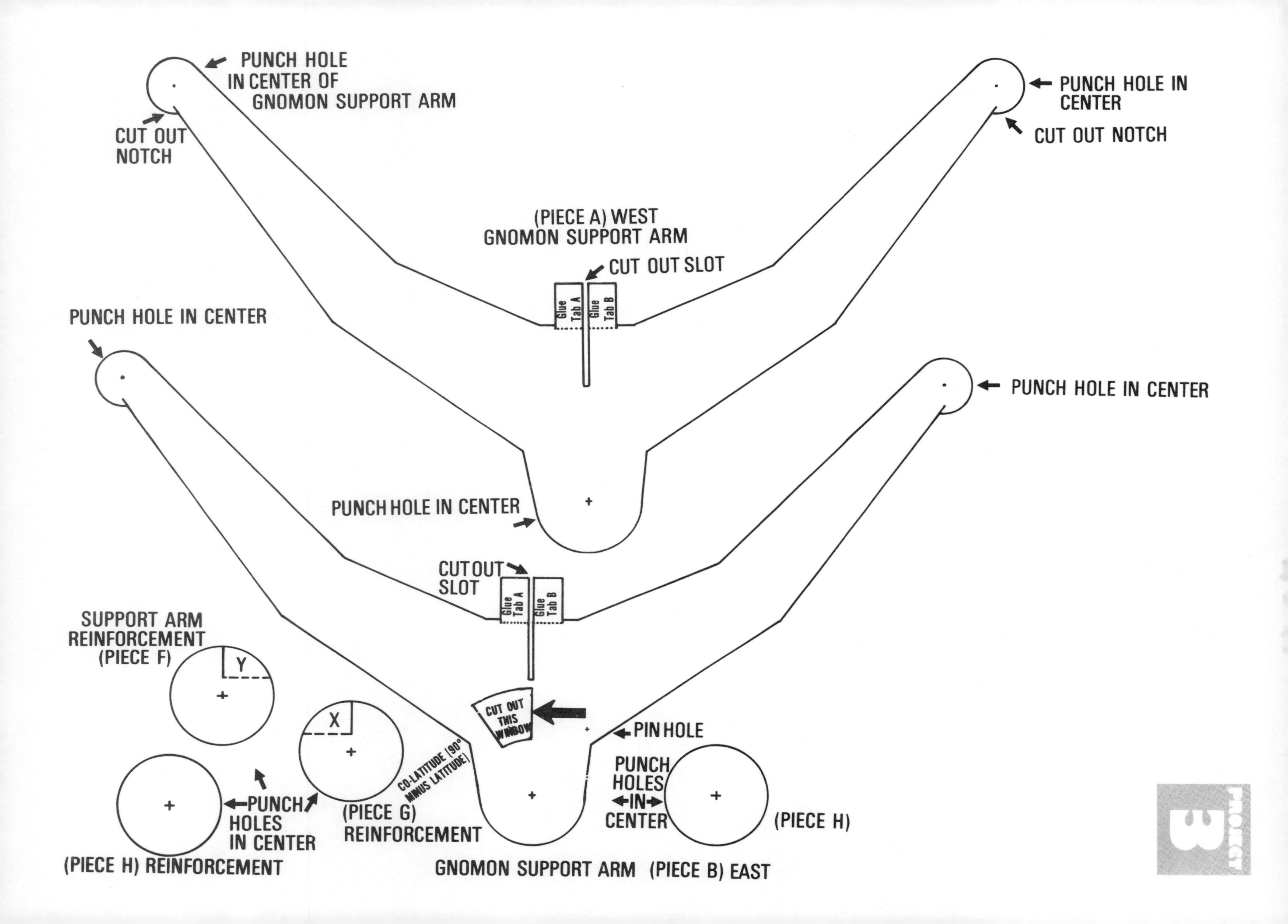
PUNCH HOLE
IN CENTER OF
GNOMON SUPPORT ARM
CUT OUT
NOTCH
PUNCH HOLE IN
CENTER
CUT OUT NOTCH
(PIECE A) WEST
GNOMON SUPPORT ARM
CUT OUT SLOT
Glue
Tab A
Glue
Tab B
PUNCH HOLE IN CENTER
PUNCH HOLE IN CENTER
PUNCH HOLE IN CENTER
CUT OUT
SLOT
Glue
Tab A
Glue
Tab B
SUPPORT ARM
REINFORCEMENT
(PIECE F)
Y
X
CUT OUT
THIS
WINDOW
PIN HOLE
CO-LATITUDE (90°
MINUS LATITUDE)
PUNCH
HOLES
IN
CENTER
PUNCH
HOLES
IN CENTER
(PIECE G)
REINFORCEMENT
(PIECE H)
(PIECE H) REINFORCEMENT
GNOMON SUPPORT ARM (PIECE B) EAST
PROJECT
3

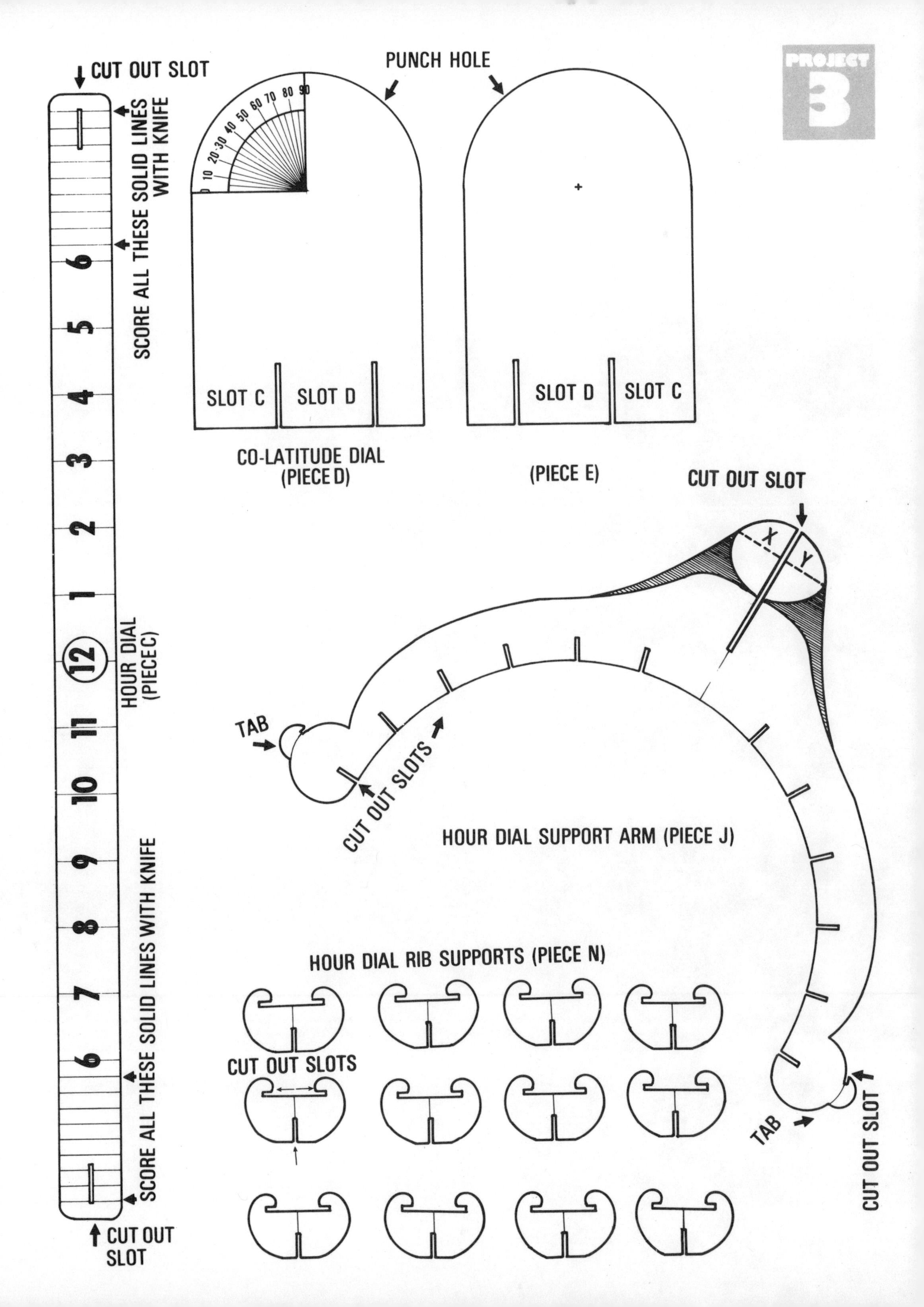
PROJECT
3
CUT OUT SLOT
SCORE ALL THESE SOLID LINES WITH KNIFE
6
5
4
3
2
1
12
HOUR DIAL (PIECE C)
11
10
9
8
7
6
SCORE ALL THESE SOLID LINES WITH KNIFE
CUT OUT SLOT
PUNCH HOLE
0 10 20 30 40 50 60 70 80 90
SLOT C
SLOT D
CO-LATITUDE DIAL (PIECE D)
SLOT D
SLOT C
(PIECE E)
CUT OUT SLOT
X
Y
TAB
CUT OUT SLOTS
HOUR DIAL SUPPORT ARM (PIECE J)
TAB
CUT OUT SLOT
HOUR DIAL RIB SUPPORTS (PIECE N)
CUT OUT SLOTS

CROSS SUNDIAL

WITH ADJUSTMENT FOR LATITUDE

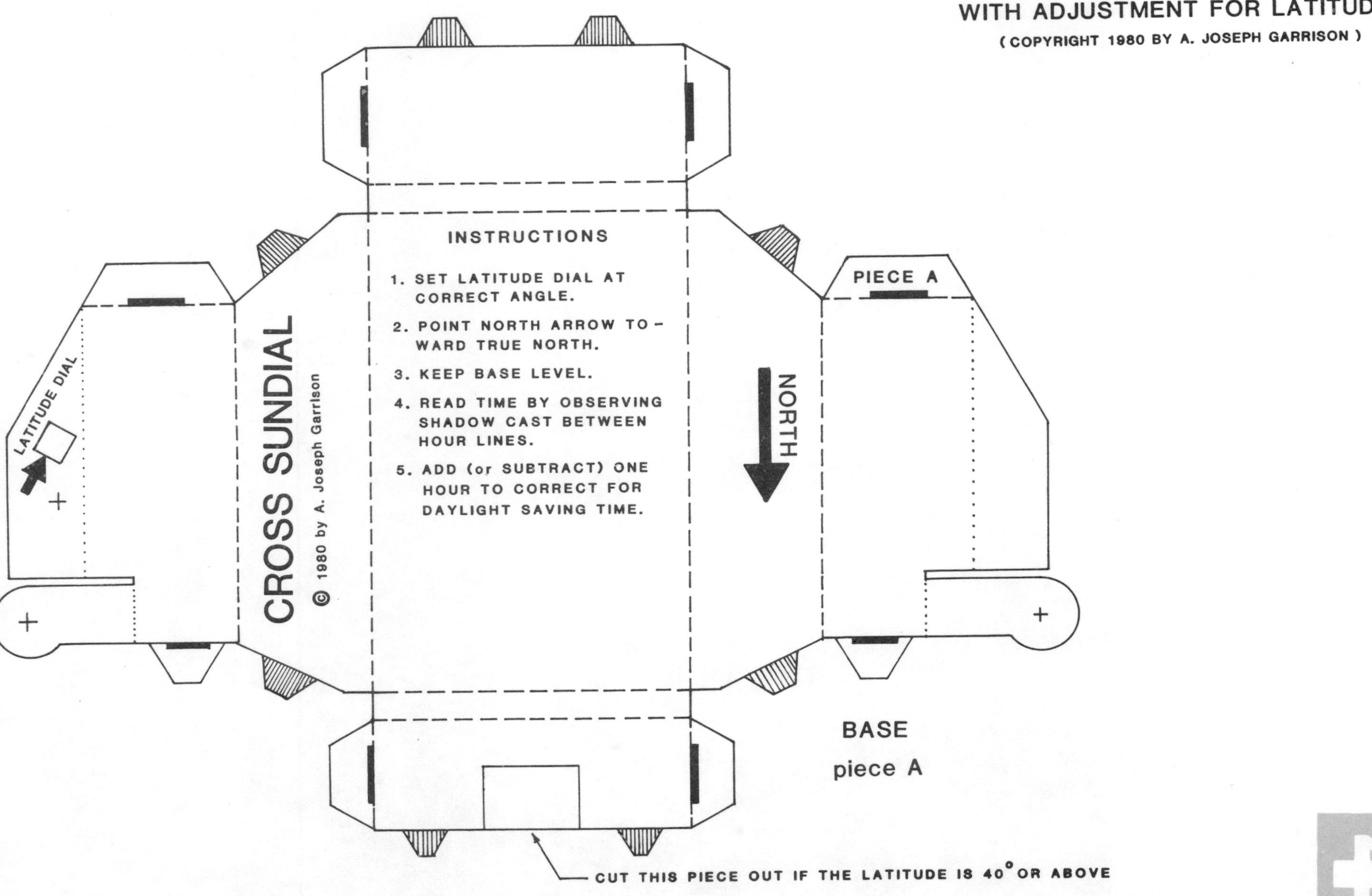

INSTRUCTIONS
1. SET LATITUDE DIAL AT CORRECT ANGLE.
2. POINT NORTH ARROW TO-WARD TRUE NORTH.
3. KEEP BASE LEVEL.
4. READ TIME BY OBSERVING SHADOW CAST BETWEEN HOUR LINES.
5. ADD (or SUBTRACT) ONE HOUR TO CORRECT FOR DAYLIGHT SAVING TIME.
CROSS SUNDIAL
© 1980 by A. Joseph Garrison
NORTH
LATITUDE DIAL
PIECE A
BASE
piece A
CUT THIS PIECE OUT IF THE LATITUDE IS 40° OR ABOVE

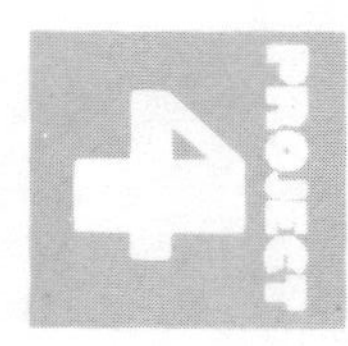

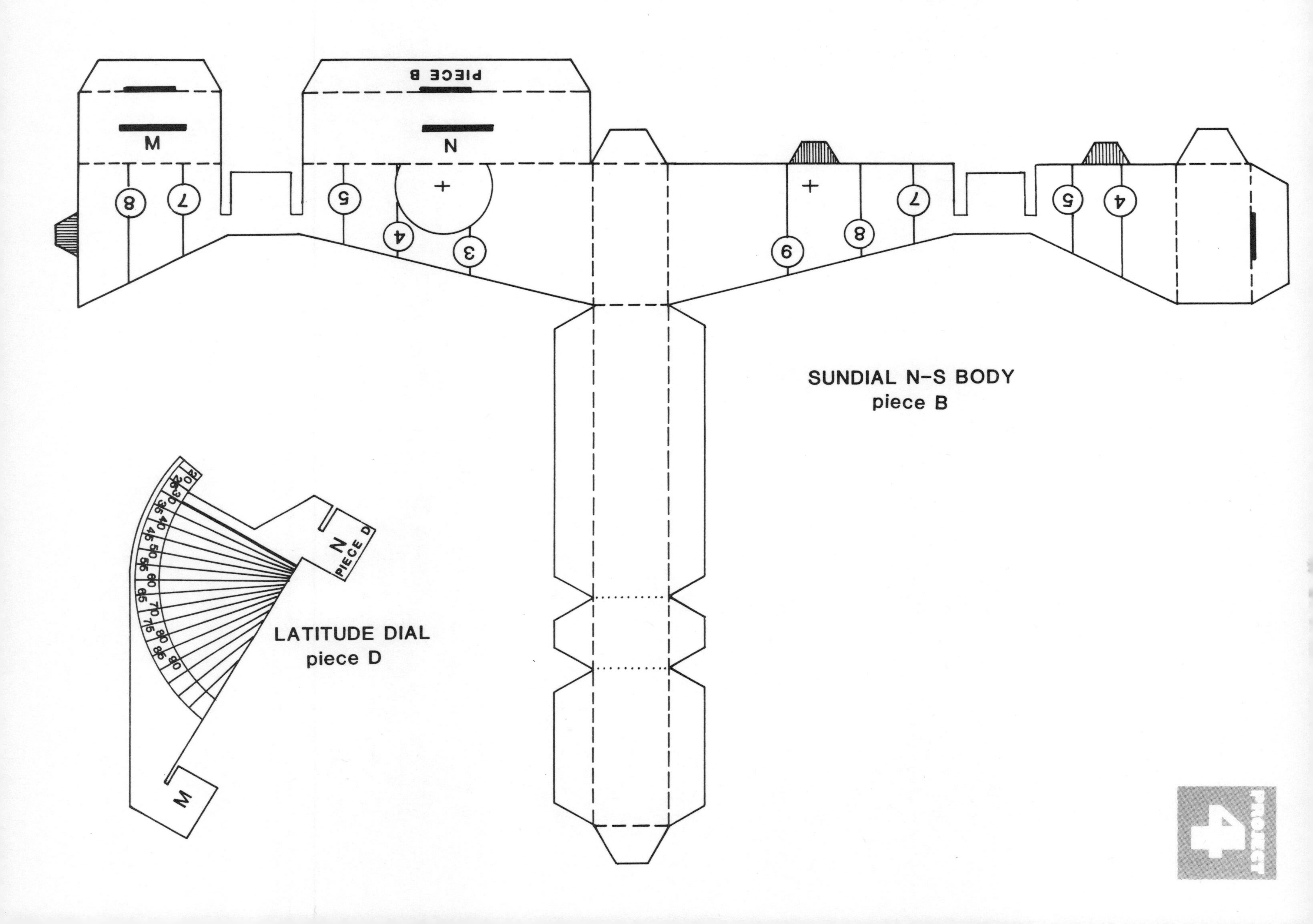

PIECE B
M
N
SUNDIAL N-S BODY
piece B
N
PIECE D
LATITUDE DIAL
piece D
M
20 25 30 35 40 45 50 55 60 65 70 75 80 85 90
PROJECT
4

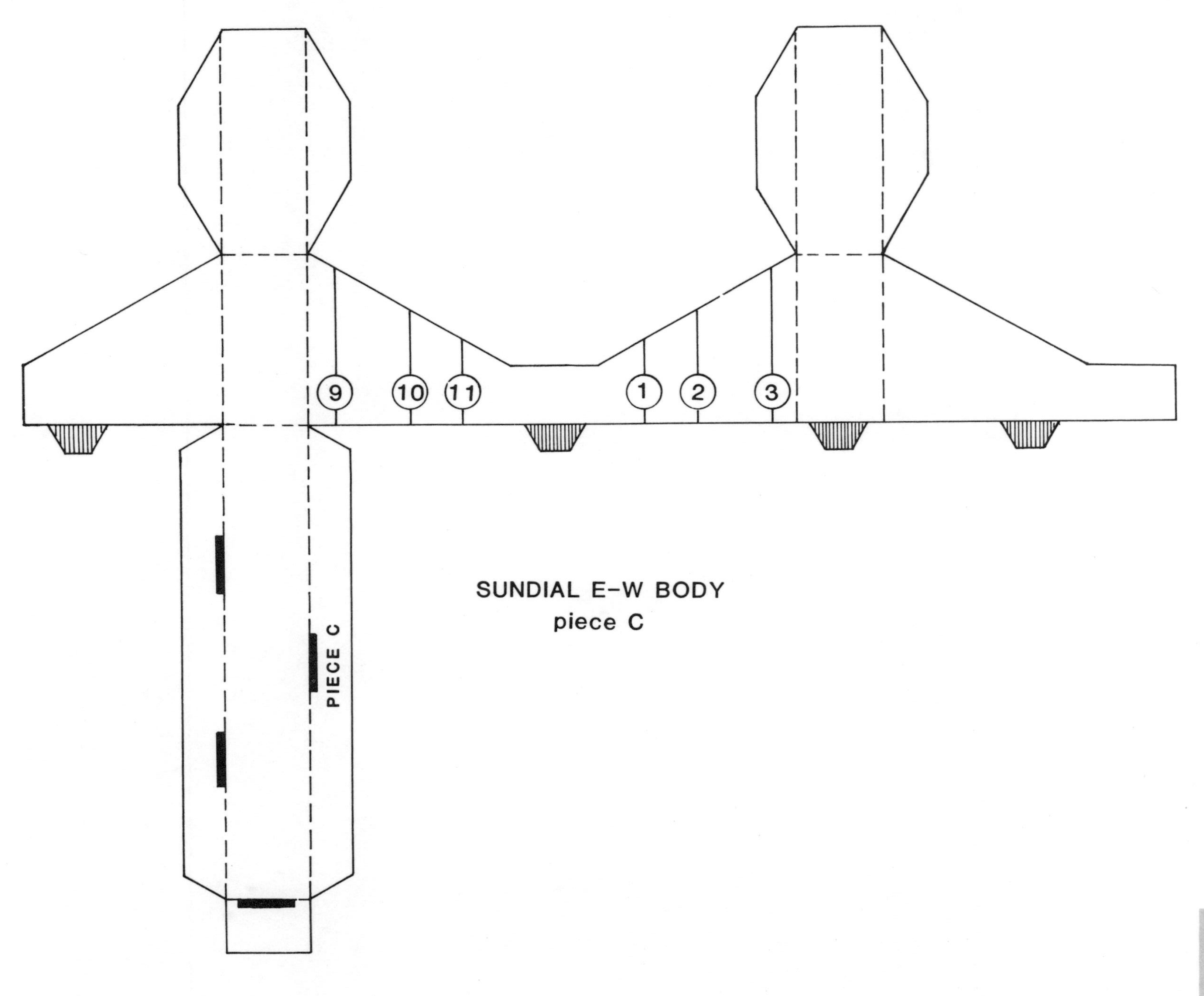
9
10
11
1
2
3
PIECE C
SUNDIAL E-W BODY
piece C

MODEL SOLAR HOUSE

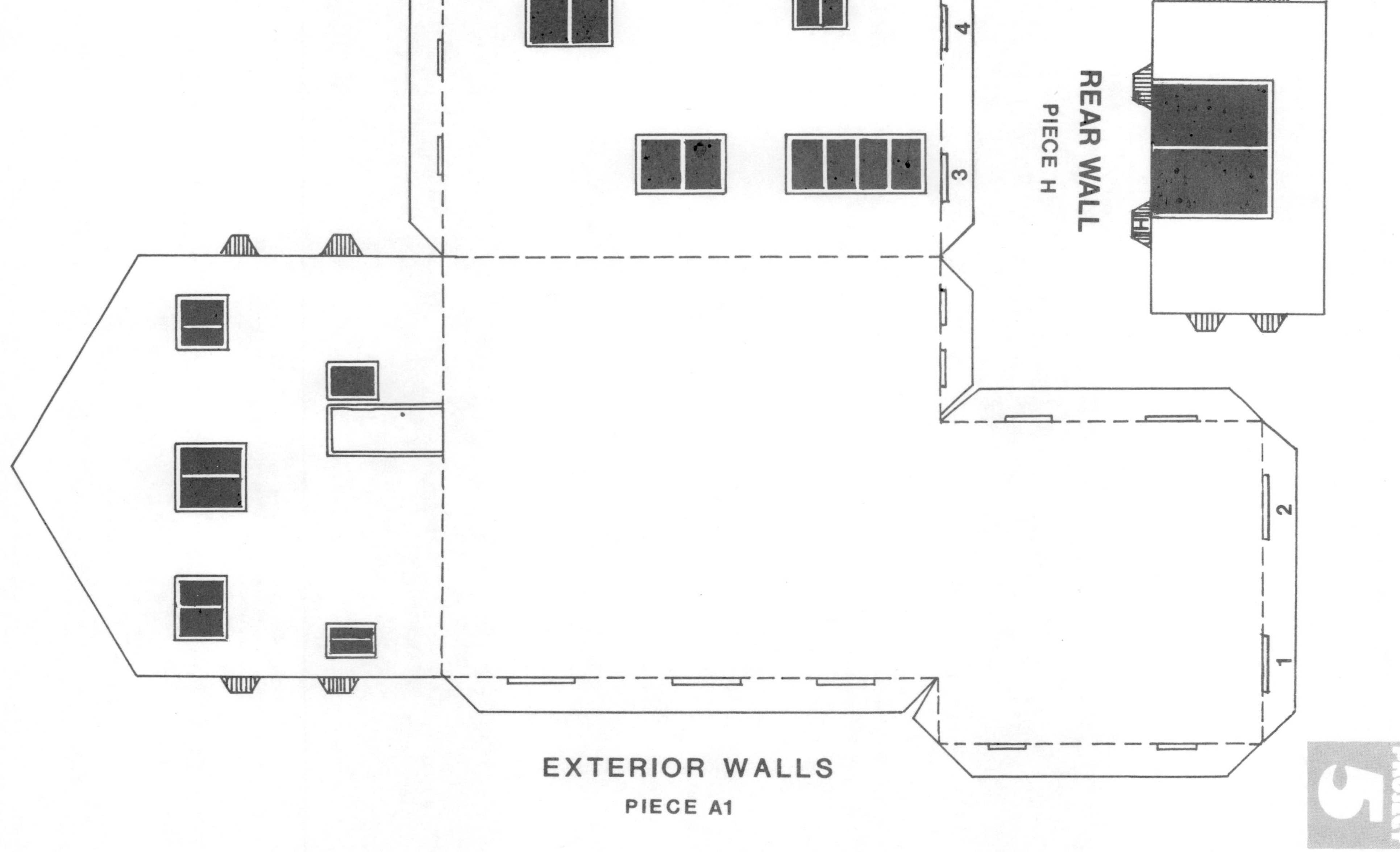

PROJECT 5

EXTERIOR WALLS

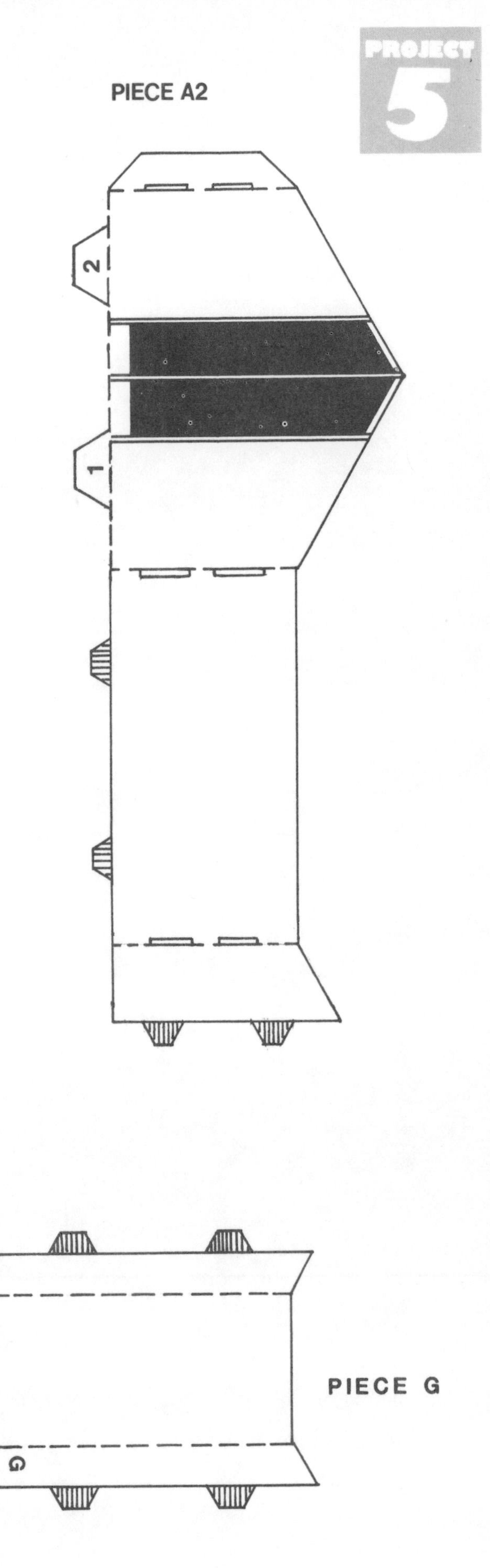

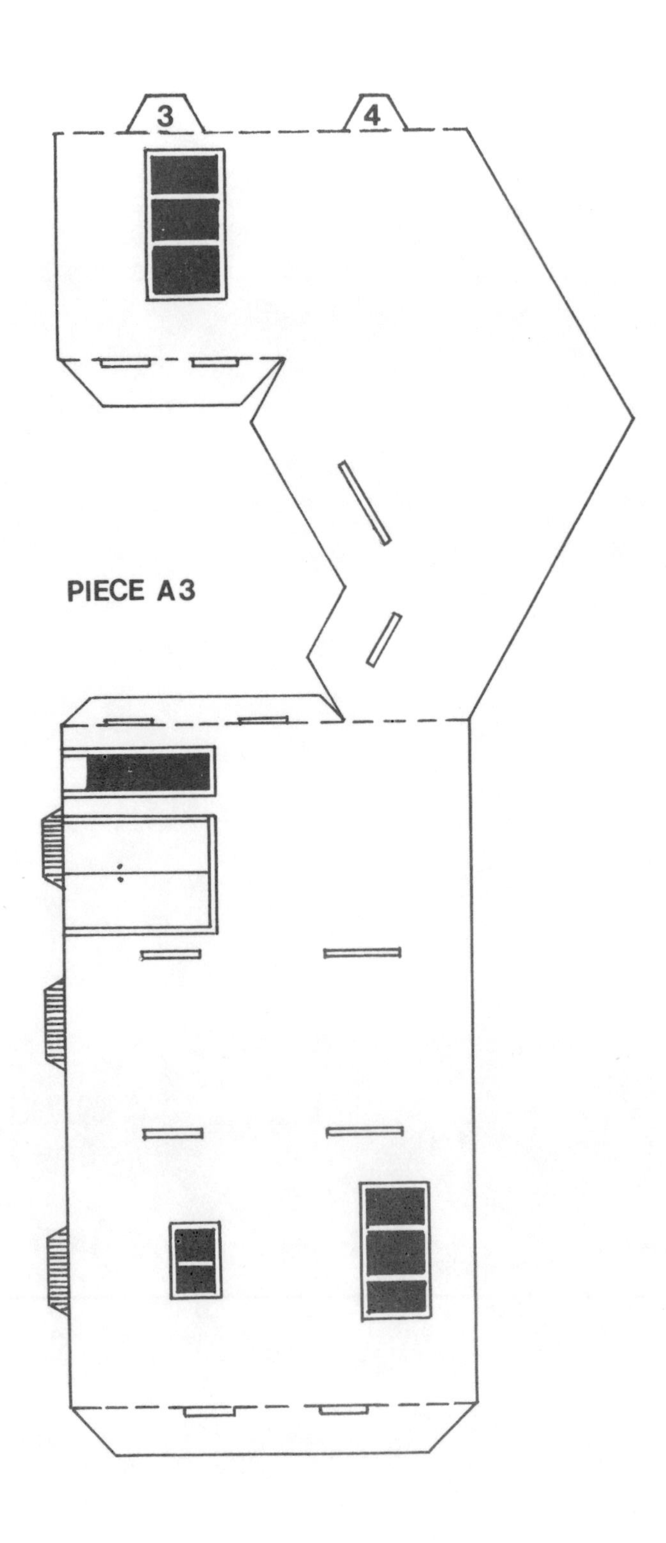

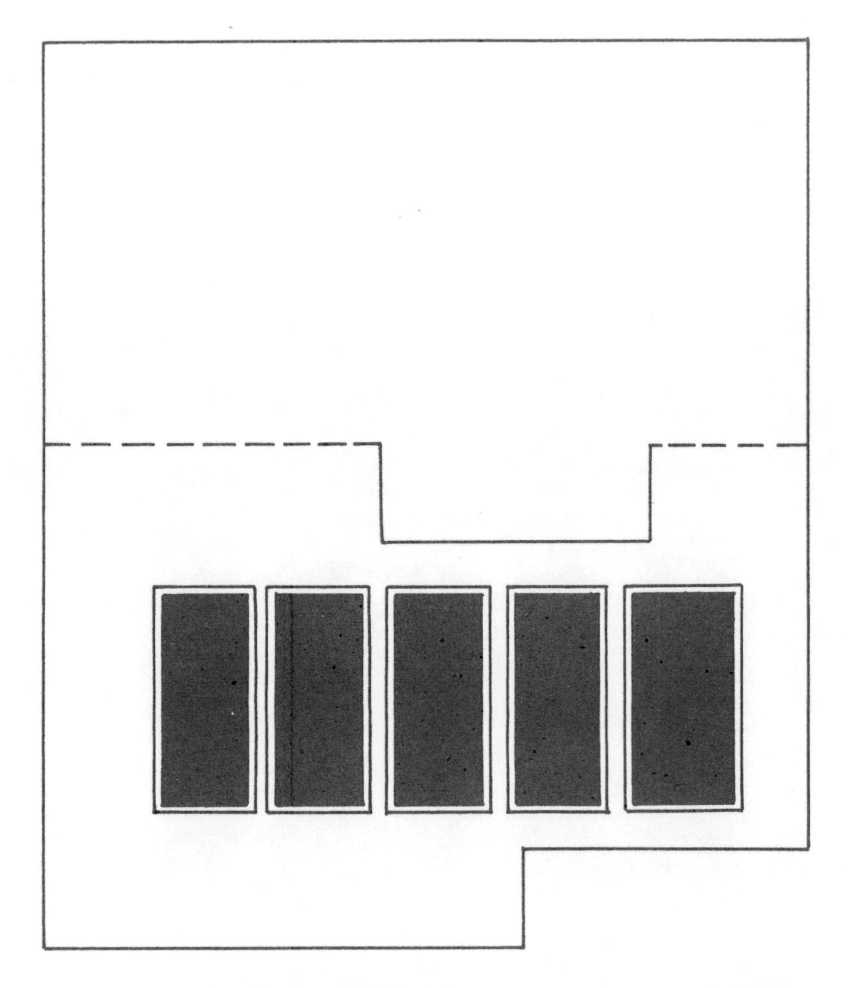

MAIN ROOF
PIECE D

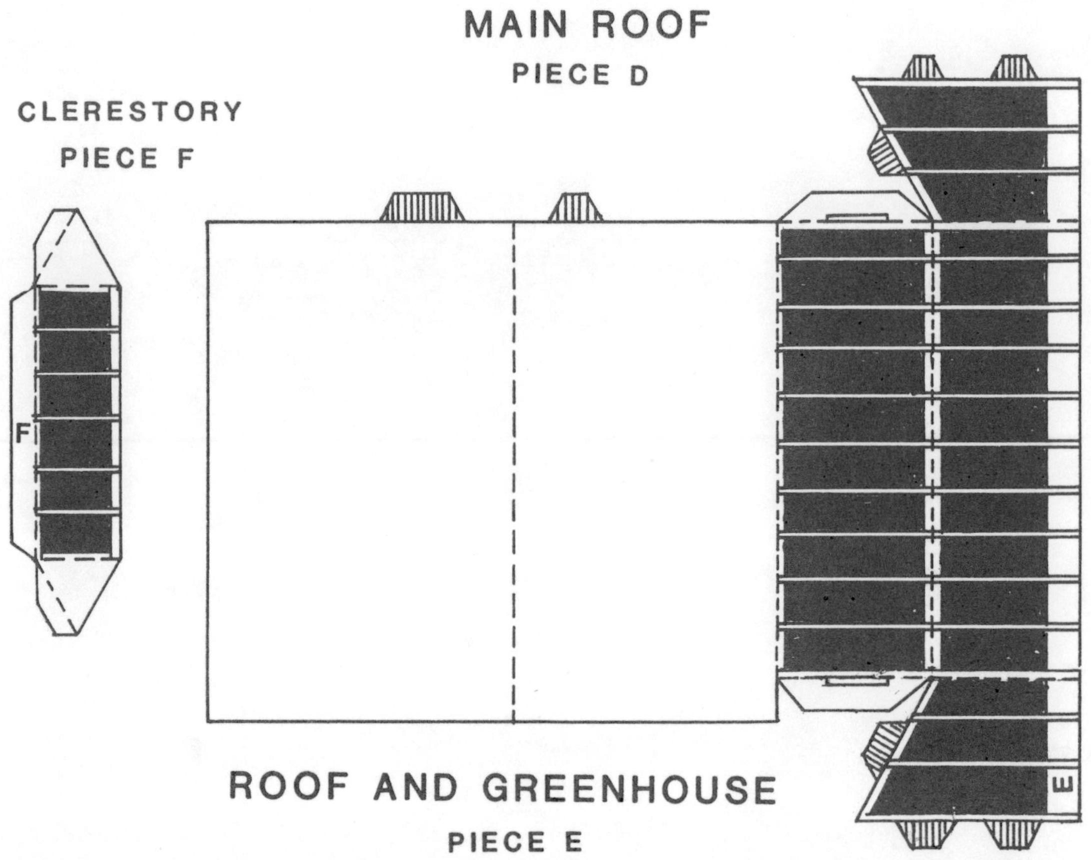

CLERESTORY
PIECE F

ROOF AND GREENHOUSE
PIECE E

PROJECT 5

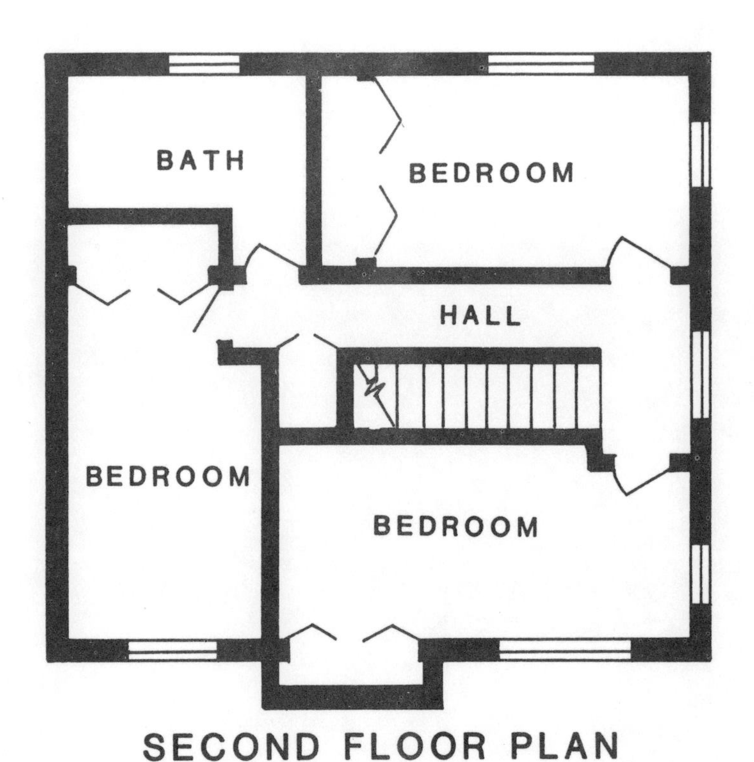

SECOND FLOOR PLAN

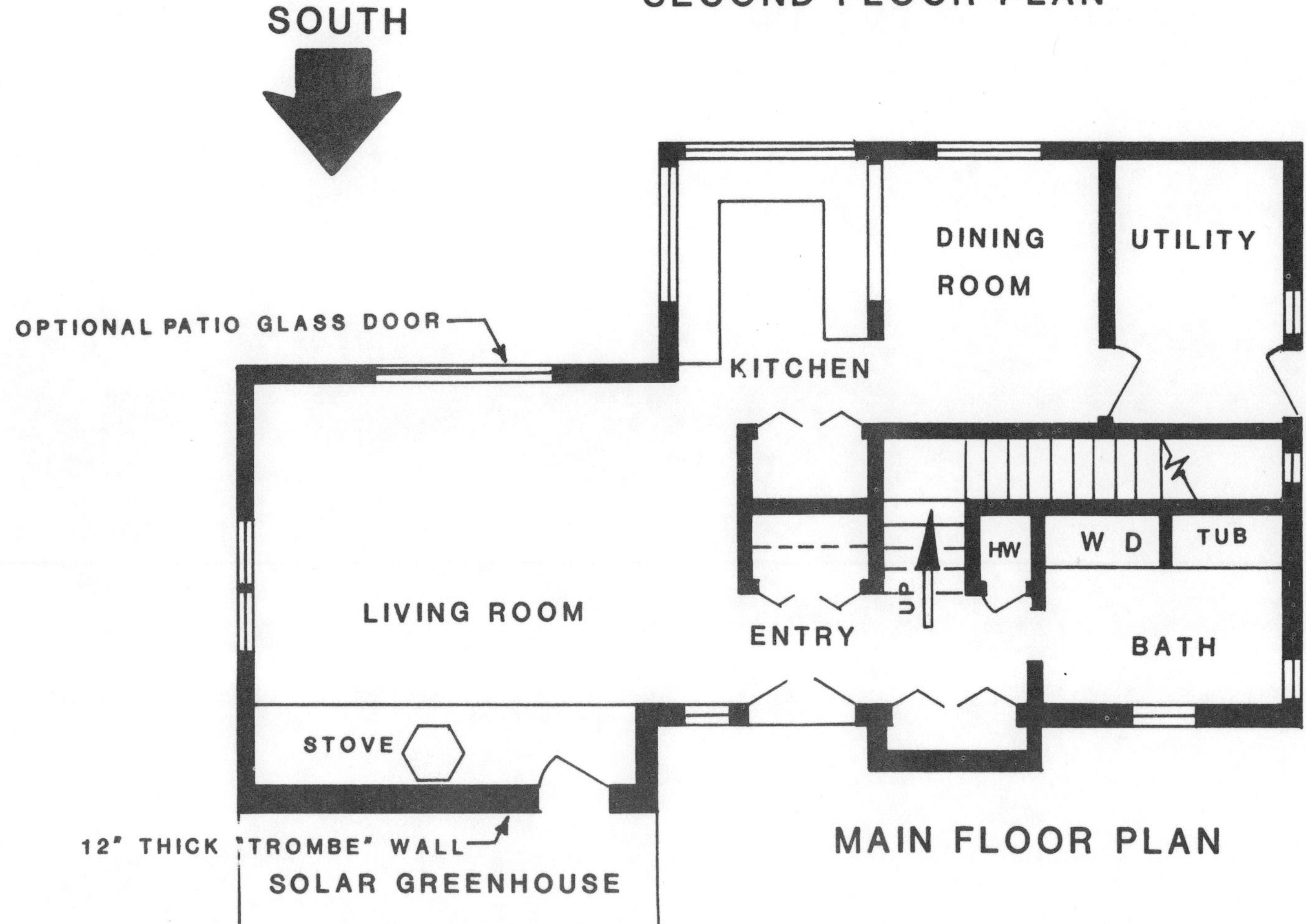

MAIN FLOOR PLAN

PROJECT
6

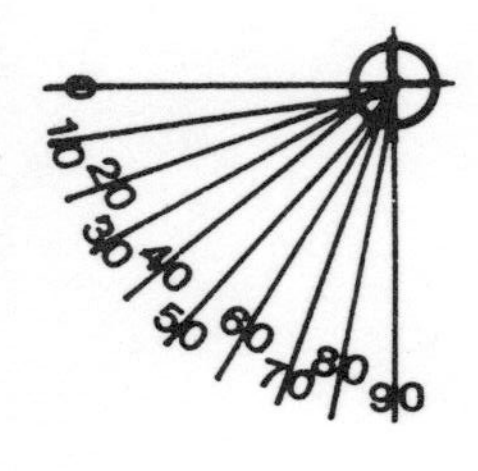

SOLAR OVEN

INSTRUCTIONS

1. PLACE FOOD ITEMS TO BE HEATED INSIDE OVEN.
2. SECURELY CLOSE LID. TAPE EDGES SHUT TO MINIMIZE HEAT LOSS.
3. FACE WINDOW SOUTH (TOWARD THE SUN).
4. ADJUST THE VERTICAL ANGLE UNTIL THE ANGLE OF THE SUN IS ATTAINED. CONTINUE READJUSTING THE ANGLE THROUGHOUT THE DAY.
5. COVER THE OVEN BODY WITH AN INSULATING MATERIAL FOR MAXIMUM HEAT GAIN.

MAIN BODY (top and side)

(piece A)

OPTIONAL: line interior of oven main body with reflective foil

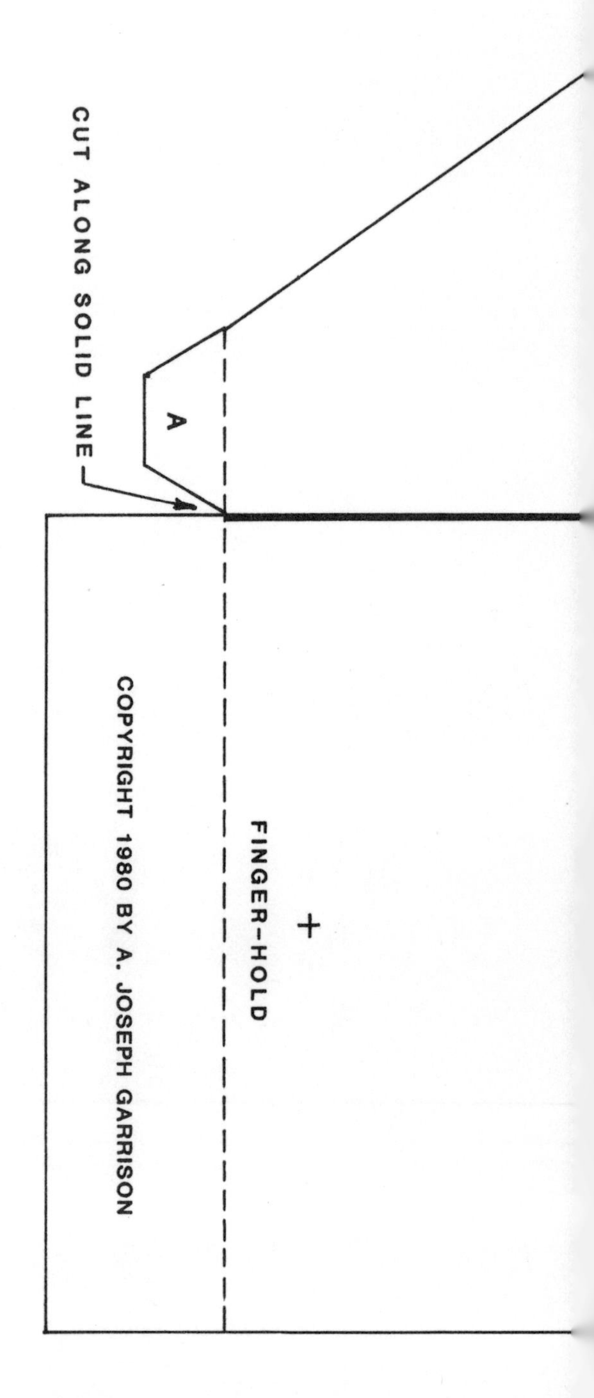

PROJECT
6

MAIN BODY (bottom and side)

(piece B)

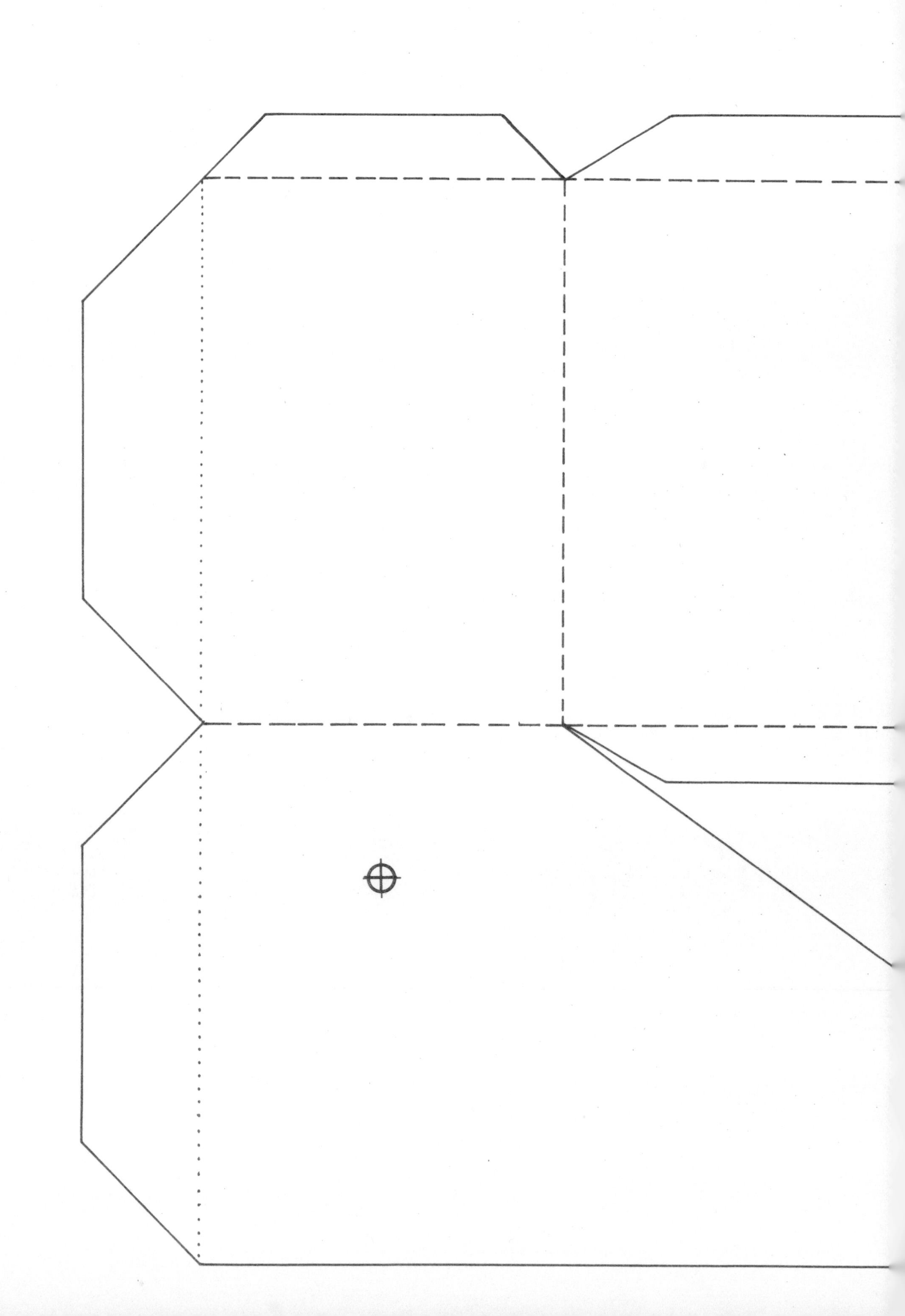

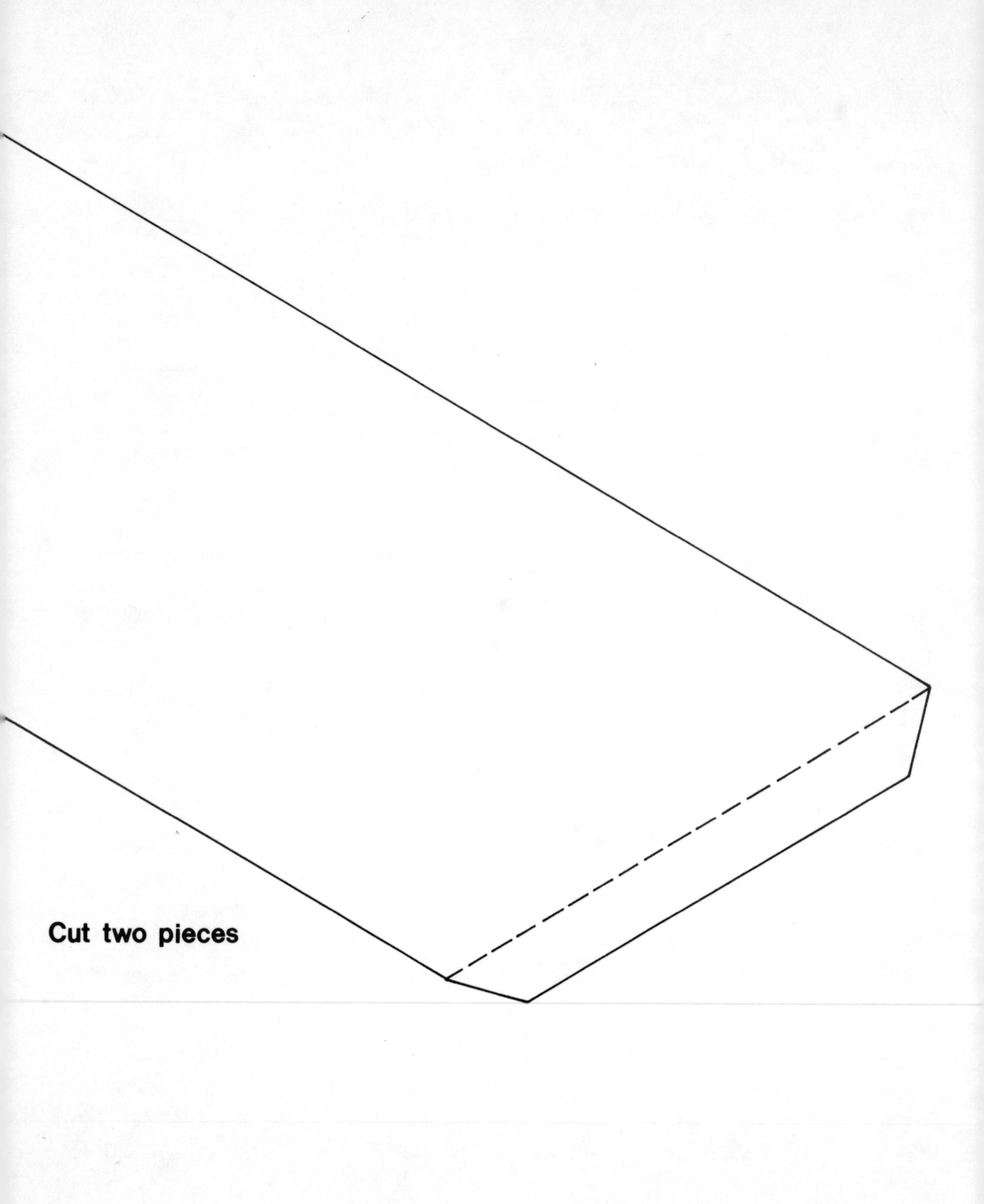
Cut two pieces

REFLECTIVE PANELS

Cover with reflective mylar or aluminum foil

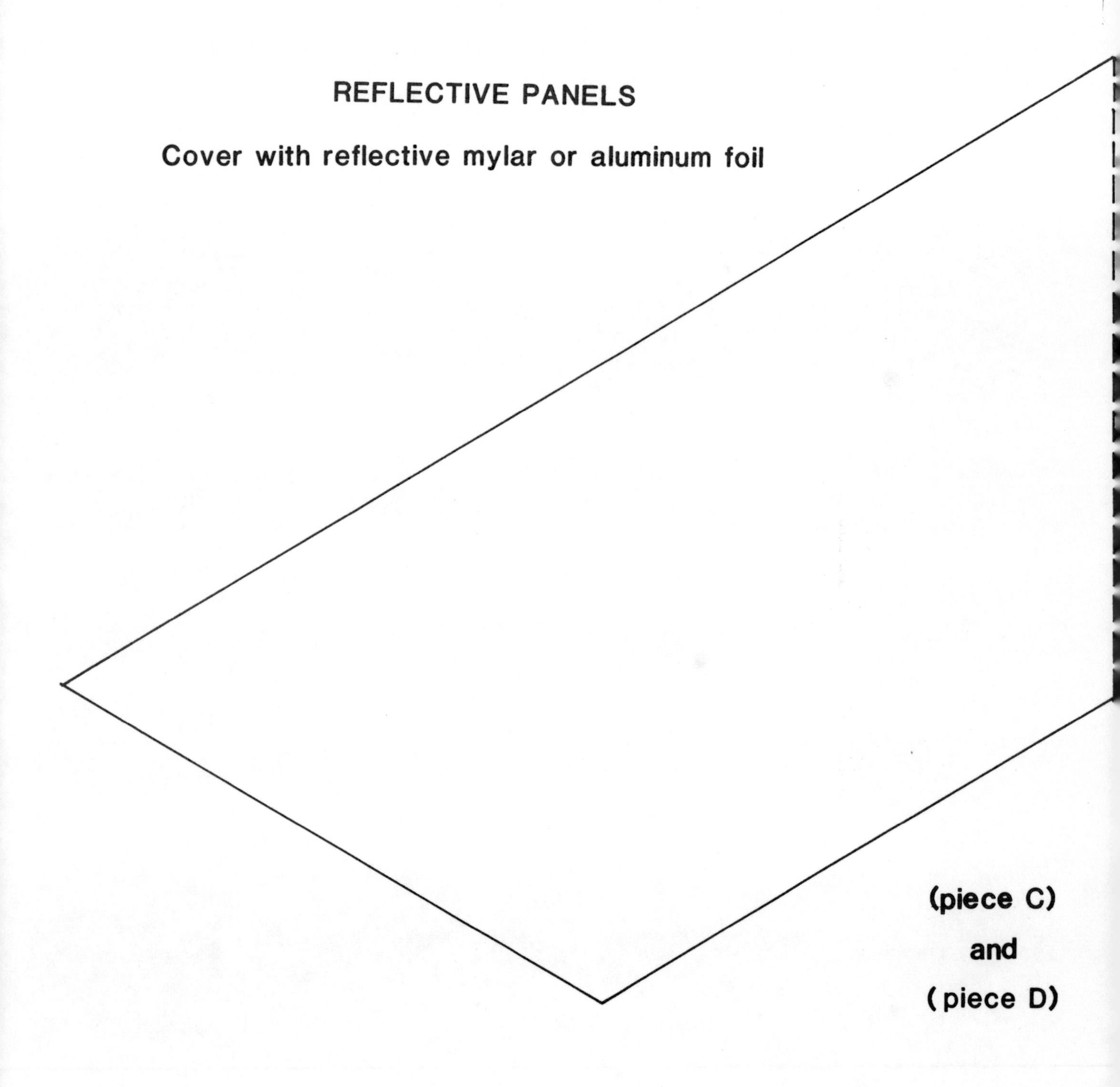

(piece C)

and

(piece D)

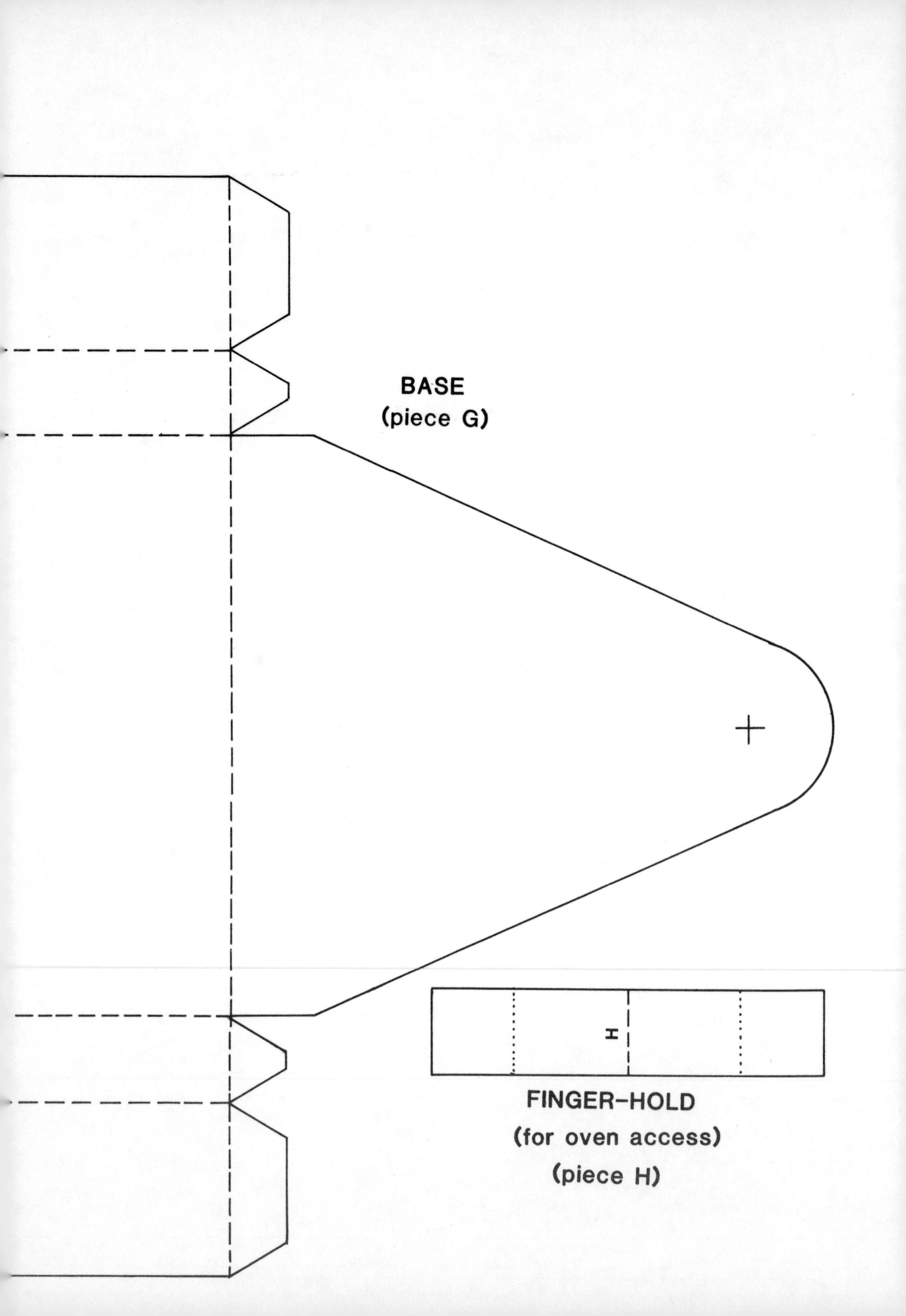
BASE
(piece G)
H
FINGER-HOLD
(for oven access)
(piece H)

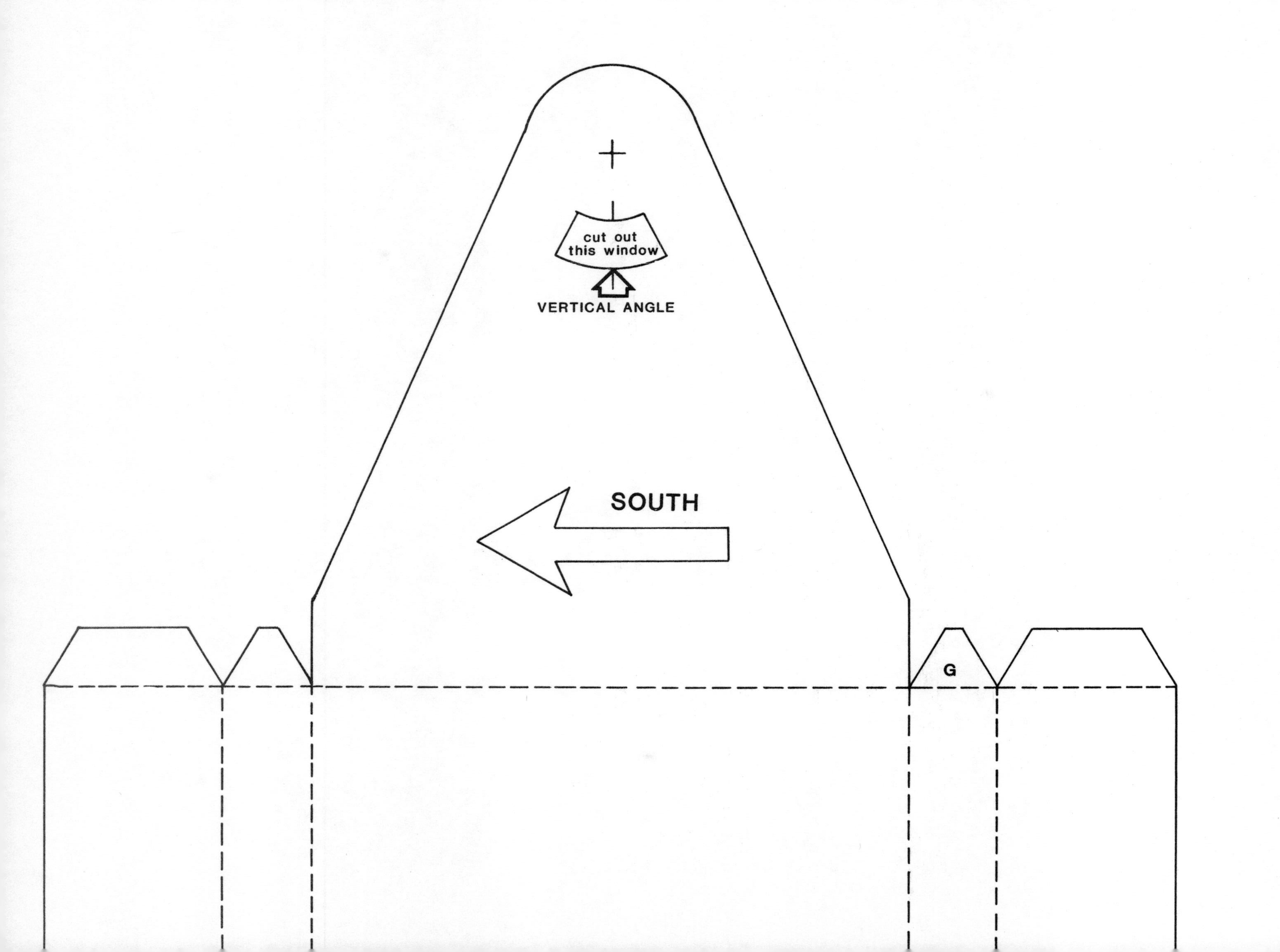
cut out
this window
VERTICAL ANGLE
SOUTH
G

PROJECT 6

WINDOW
(cut this space out)

glue "Brown-In-Bag" oven basting bag onto window frame

IF AVAILABLE, USE CLEAR MYLAR (PREFERRED)

E

(piece E)

UPPER LID CLASP
PIECE J

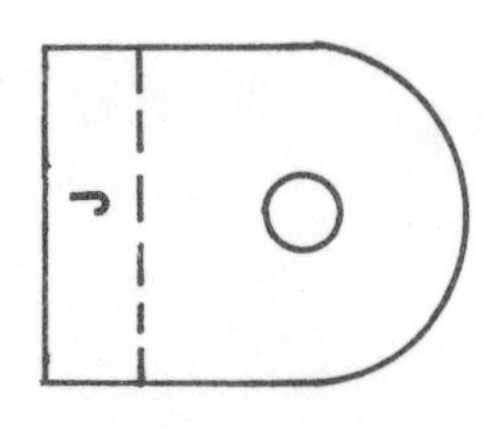

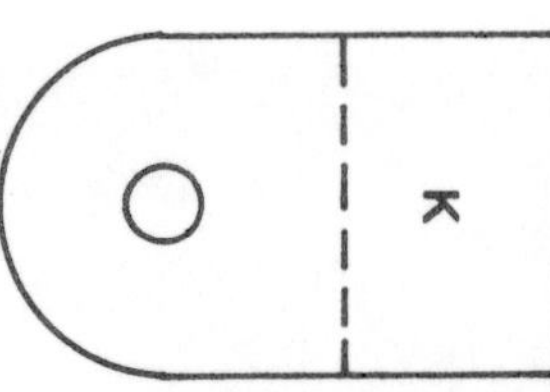

LOWER LID CLASP
PIECE K

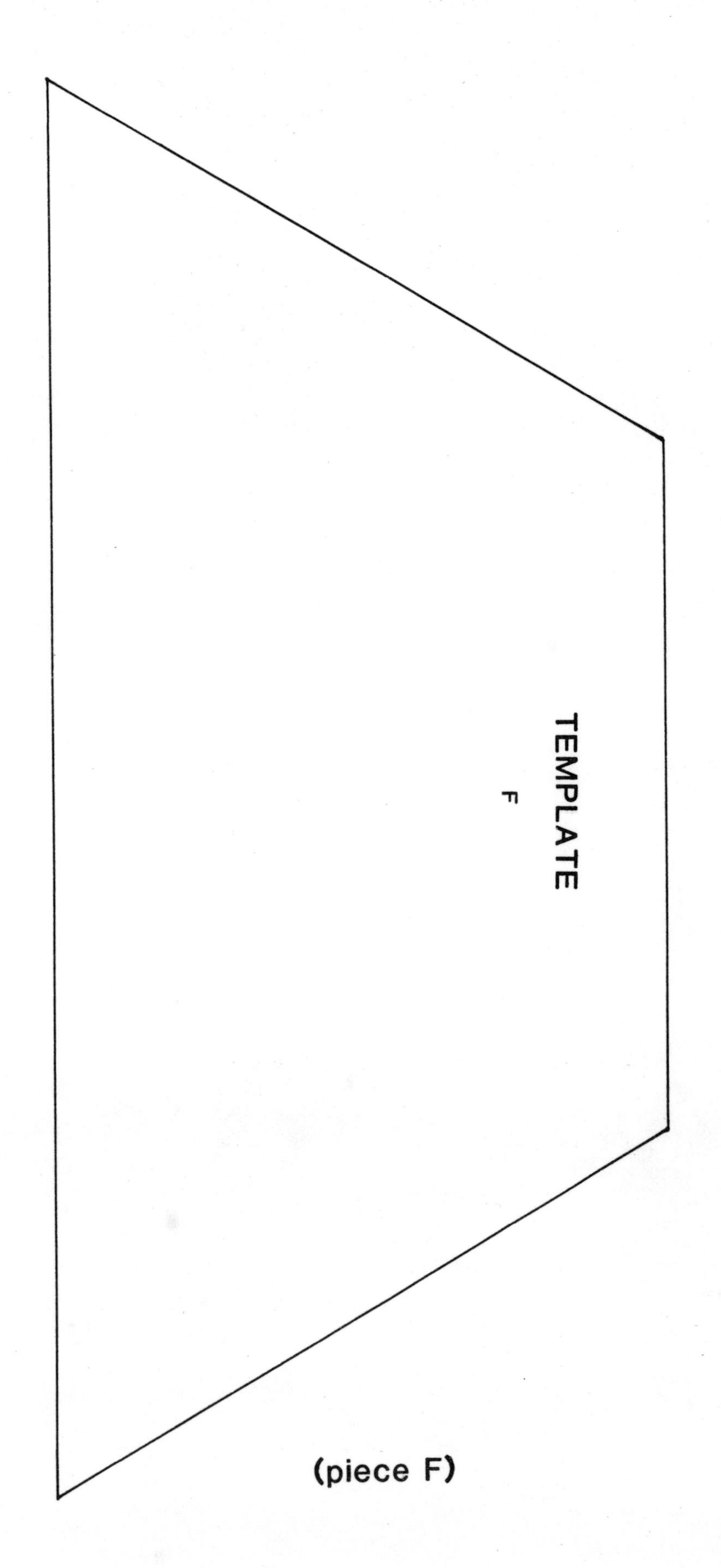

(piece F)

SOLAR
FOOD DEHYDRATOR

TOP COVER ASSEMBLY (SIDE)
PIECE D

D

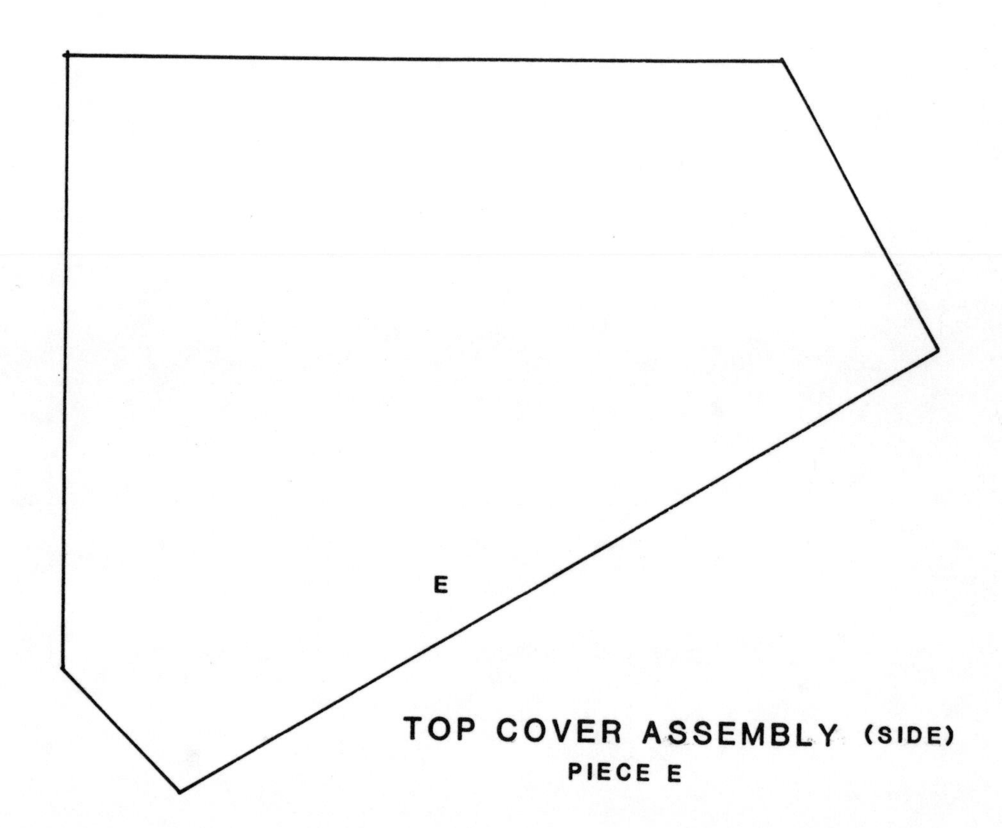

A

FRONT

(TRANSPARENT COVER)

PIECE A

COVER WINDOW WITH CLEAR SELF-BASTING OVEN BAG AND GLUE ALL EDGES AIR-TIGHT
(IF AVAILABLE, CLEAR MYLAR IS PREFERRED)

CUT OUT WINDOW

USE PAPER PUNCH TO CUT OUT HOLES

FRONT FRAME

(TRIM FOR TRANSPARENT COVER)

PIECE B

CUT OUT WINDOW →

B

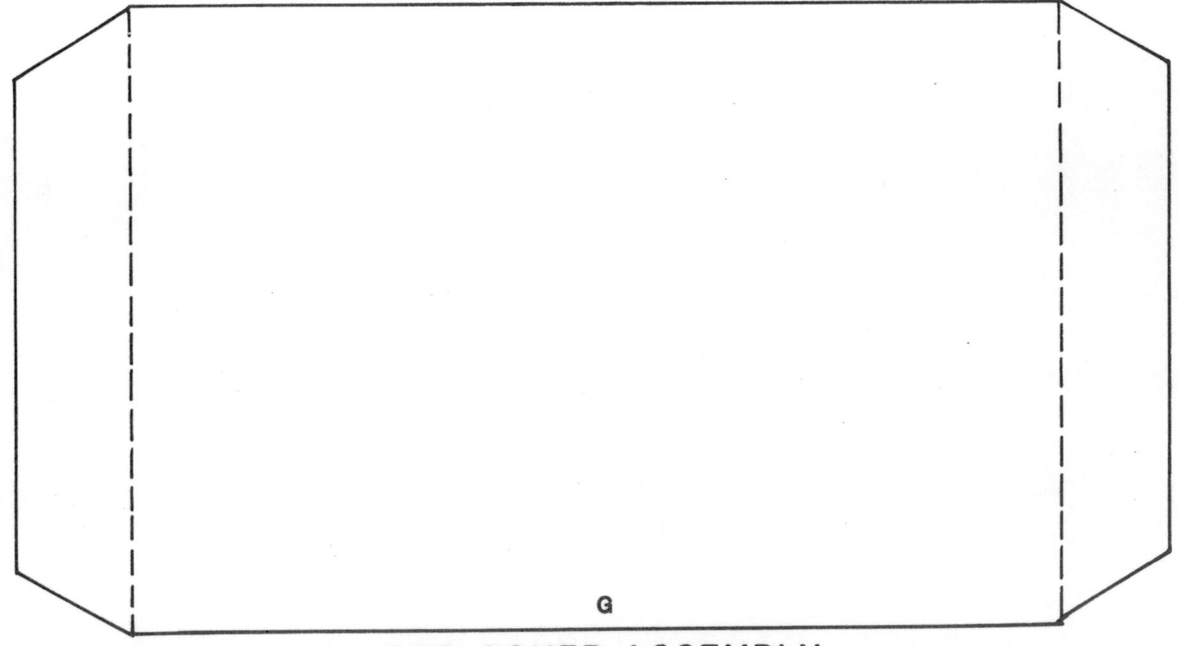

TOP COVER ASSEMBLY

(BACK INTERIOR) PIECE G

BASE

PIECE C

INSTRUCTIONS

1. WASH AND SLICE FRUIT AND VEGETABLES INTO THIN SHAPES.
2. PLACE FOOD ON METAL RODS INSIDE BOX. CLOSE LID.
3. ORIENT "SOUTH" ARROW TOWARD TRUE (CELESTIAL) SOUTH FOR MAXIMUM SOLAR EXPOSURE.
4. MAKE SURE AIR VENTS ARE NOT OBSTRUCTED; FREE CIRCULATION OF AIR IS NECESSARY.
5. TRY TO MAINTAIN A TEMPERATURE OF BETWEEN 130° F. AND 140° F. FOR BEST RESULTS. TOO HIGH A TEMPERATURE WILL COOK THE FOOD.
6. PROTECT THE FOOD FROM INSECTS. VENTS MAY BE COVERED WITH CHEESE CLOTH, IF NECESSARY.
7. TURN THE DRYING FOOD TWICE A DAY.
8. TAKE THE FOOD INSIDE AT NIGHT OR DURING CLOUDY WEATHER.
9. TO PREVENT INSECT EGGS FROM BEING HATCHED, THE DRIED FOOD CAN BE HEATED TO 180o F FOR 5 TO 10 MINUTES.
10. FOOD MAY BE BLANCHED (BOILED) FOR A FEW MINUTES BEFORE DEHYDRATING. THIS IS A COMMON PRACTICE TO ELIMINATE BAD ODORS AND DESTRUCTIVE CHEMICAL CHANGES.

APPROXIMATELY 10 POUNDS OF FRESH FRUIT AND VEGETABLES WILL REDUCE TO 1 POUND AFTER DEHYDRATION.

THE DEHYDRATOR MAY BE COVERED WITH INSULATION IF HIGHER TEMPERATURES ARE NEEDED. (CAUTION: DO NOT COVER VENTS)

F

TOP COVER ASSEMBLY

PIECE F

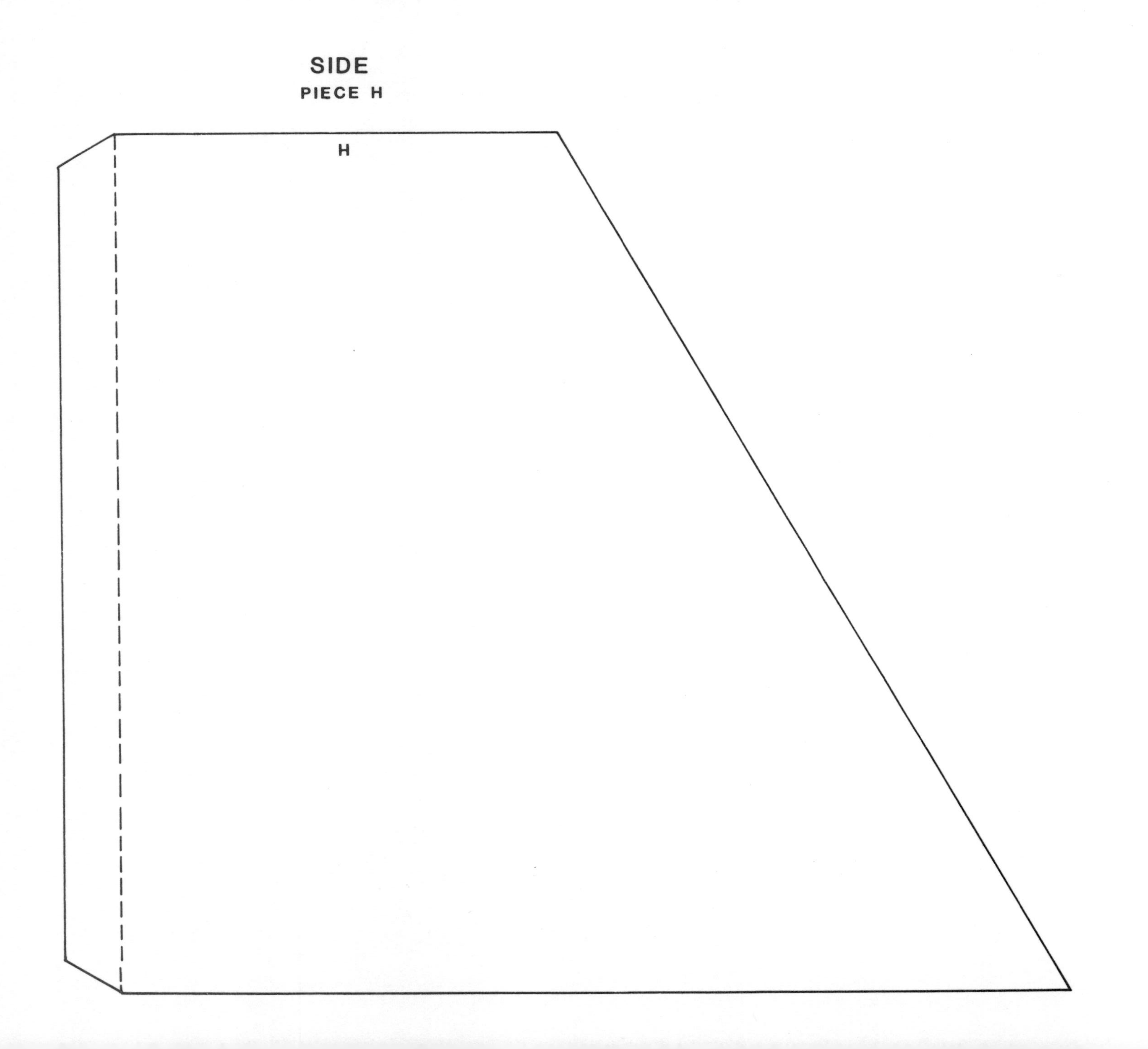
SIDE
PIECE H
H

J

SOUTH

SIDE
PIECE J

COVER WITH REFLECTIVE
MYLAR OR ALUMINUM FOIL

M

REFLECTIVE FIN
PIECE M

L

BACK
PIECE L

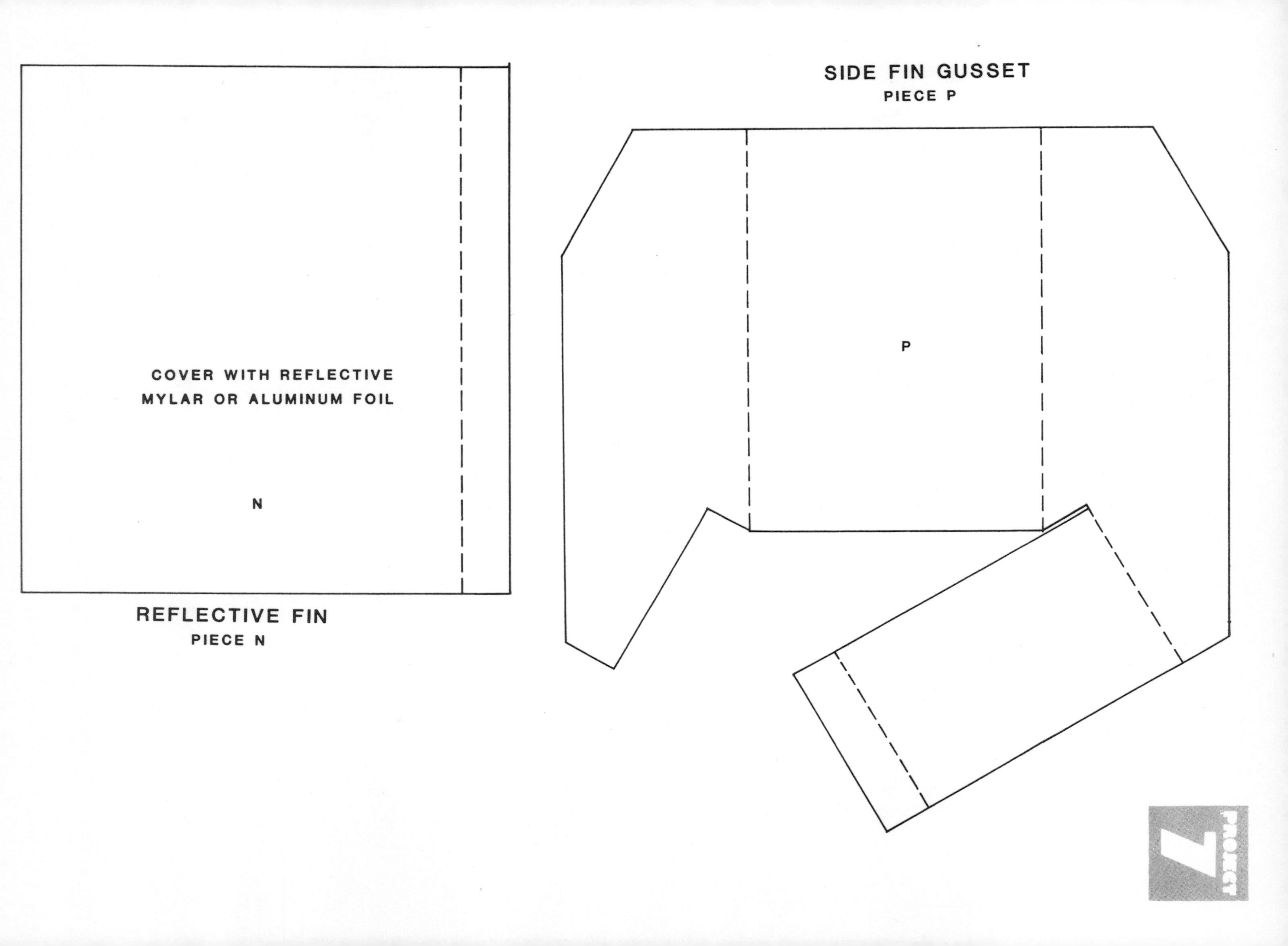

SIDE FIN GUSSET
PIECE P
P
COVER WITH REFLECTIVE
MYLAR OR ALUMINUM FOIL
N
REFLECTIVE FIN
PIECE N
PROJECT
7

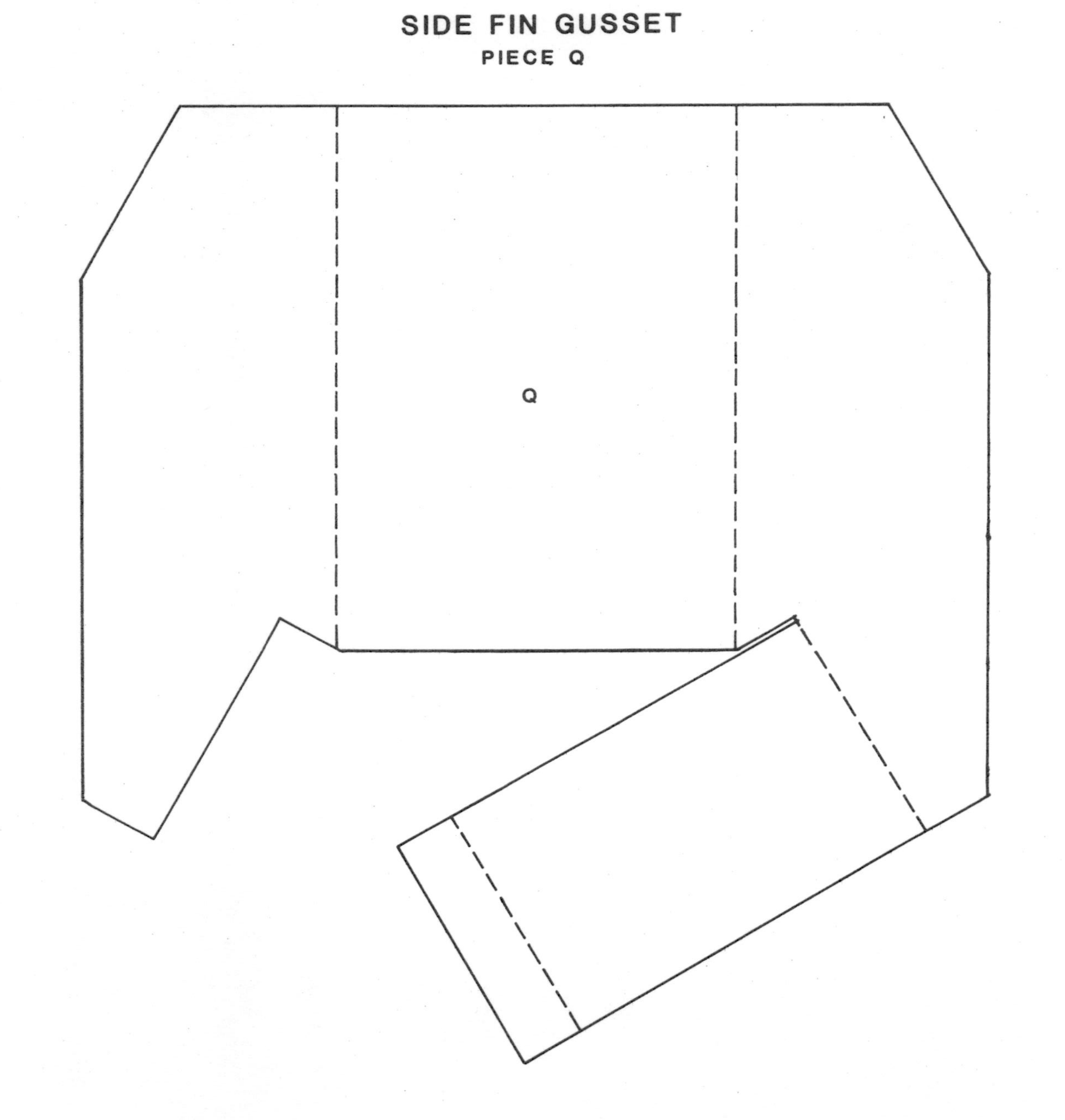
SIDE FIN GUSSET
PIECE Q
Q

INTERIOR SUPPORT FRAME (SIDE)

PIECE S

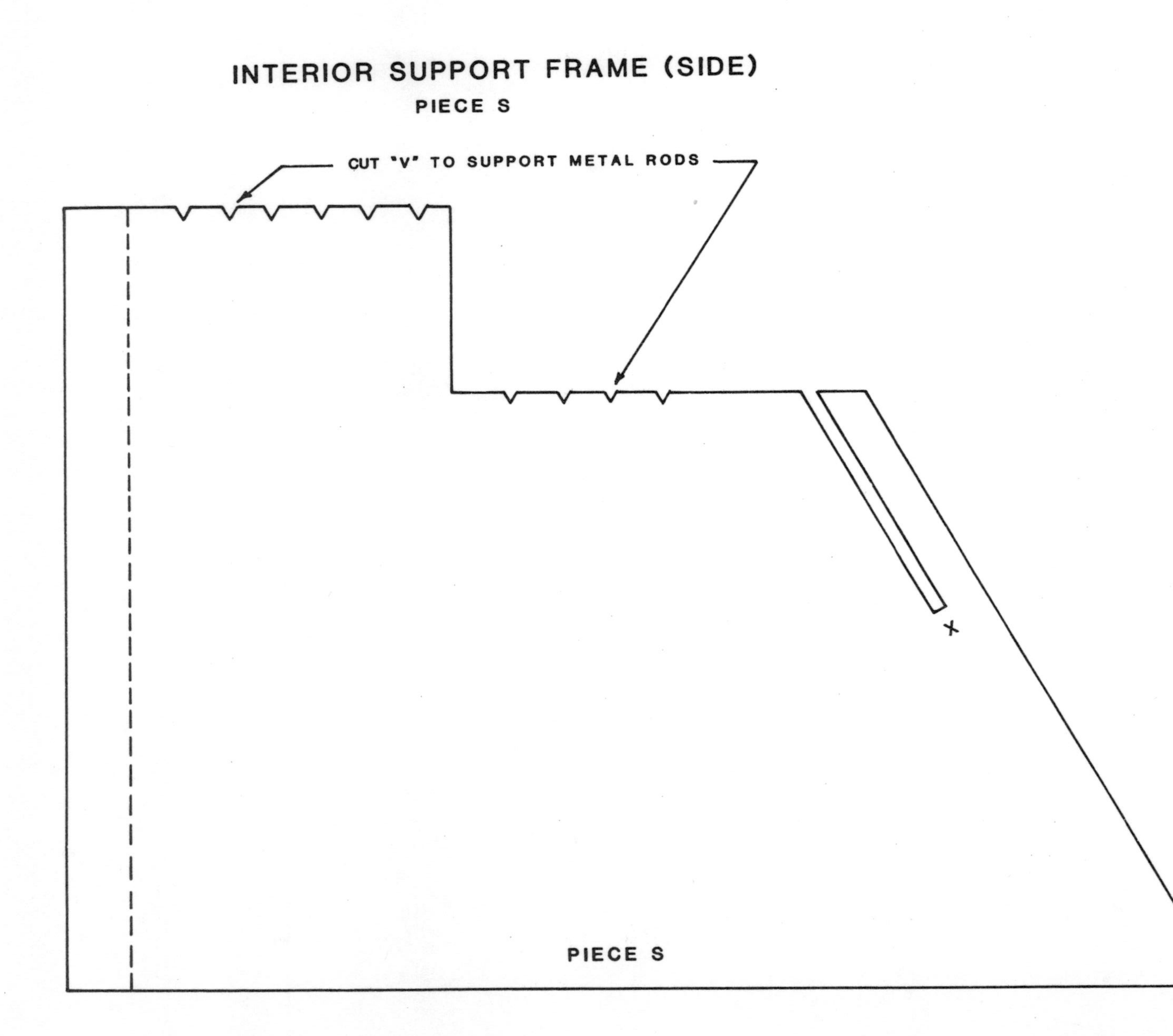

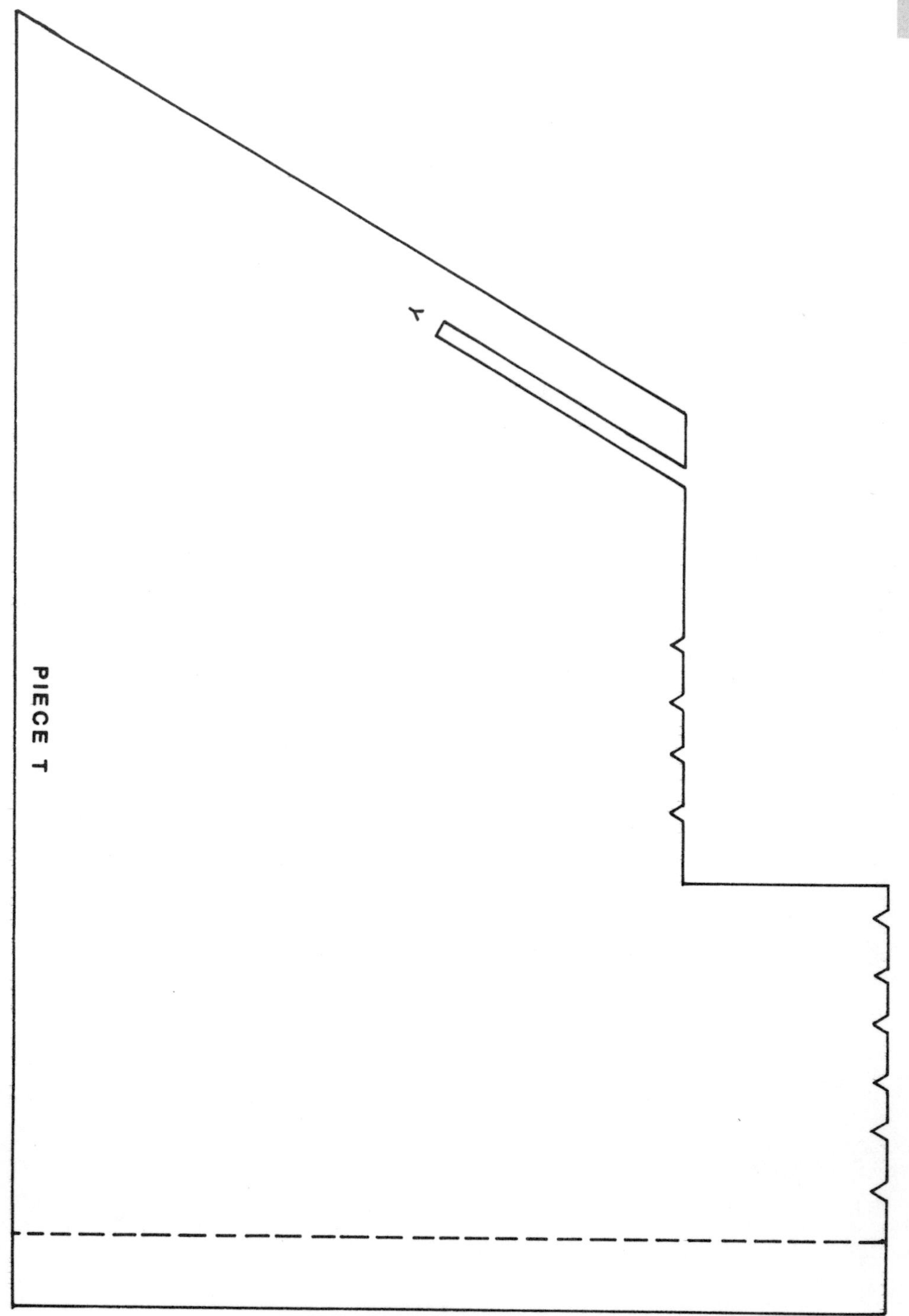

INTERIOR FRAME SUPPORT (SIDE)

PIECE T

INTERIOR FRAME SUPPORT (BACK)

PIECE U

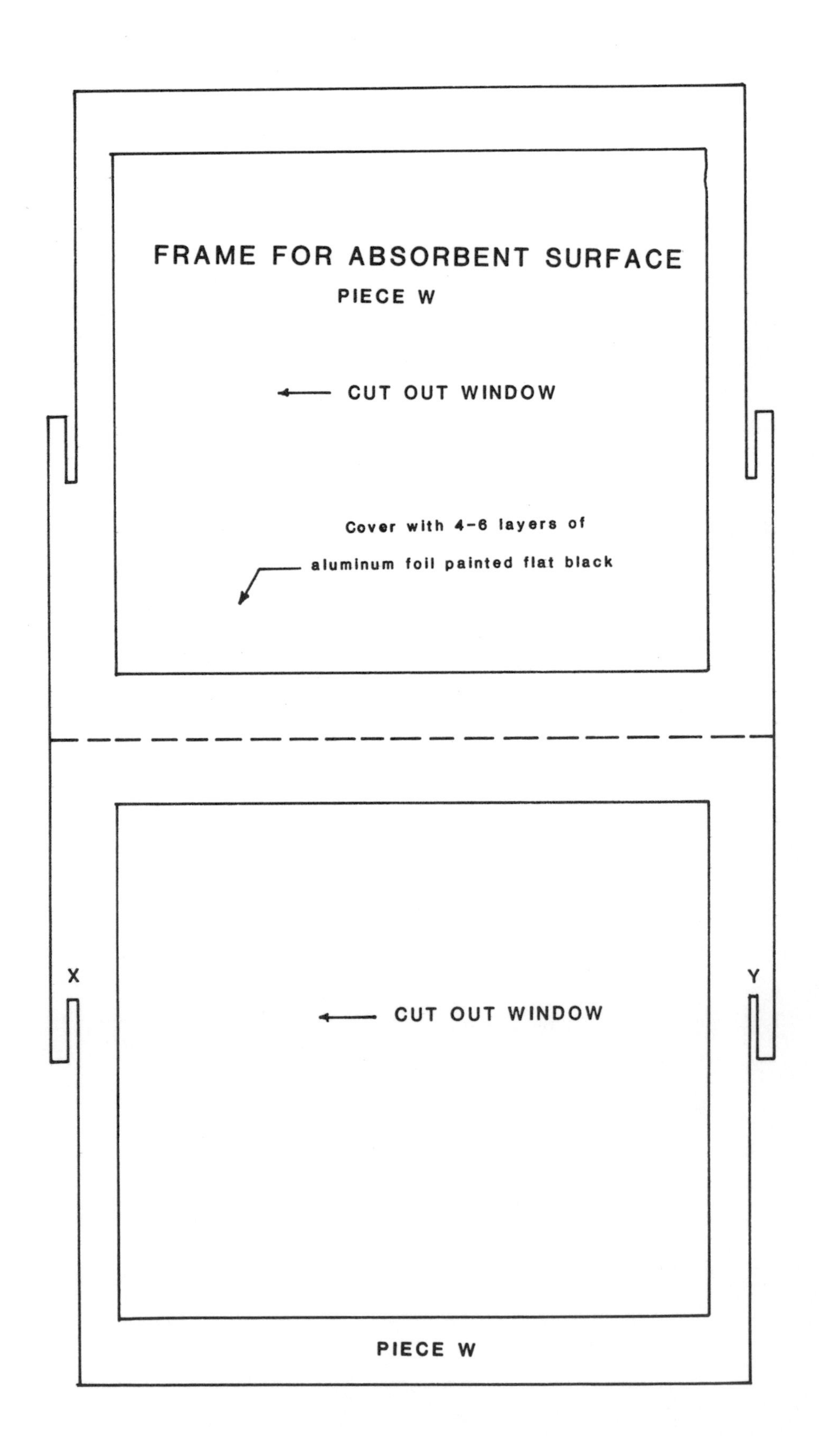
FRAME FOR ABSORBENT SURFACE
PIECE W
CUT OUT WINDOW
Cover with 4-6 layers of aluminum foil painted flat black
X
Y
CUT OUT WINDOW
PIECE W

PROJECT
8

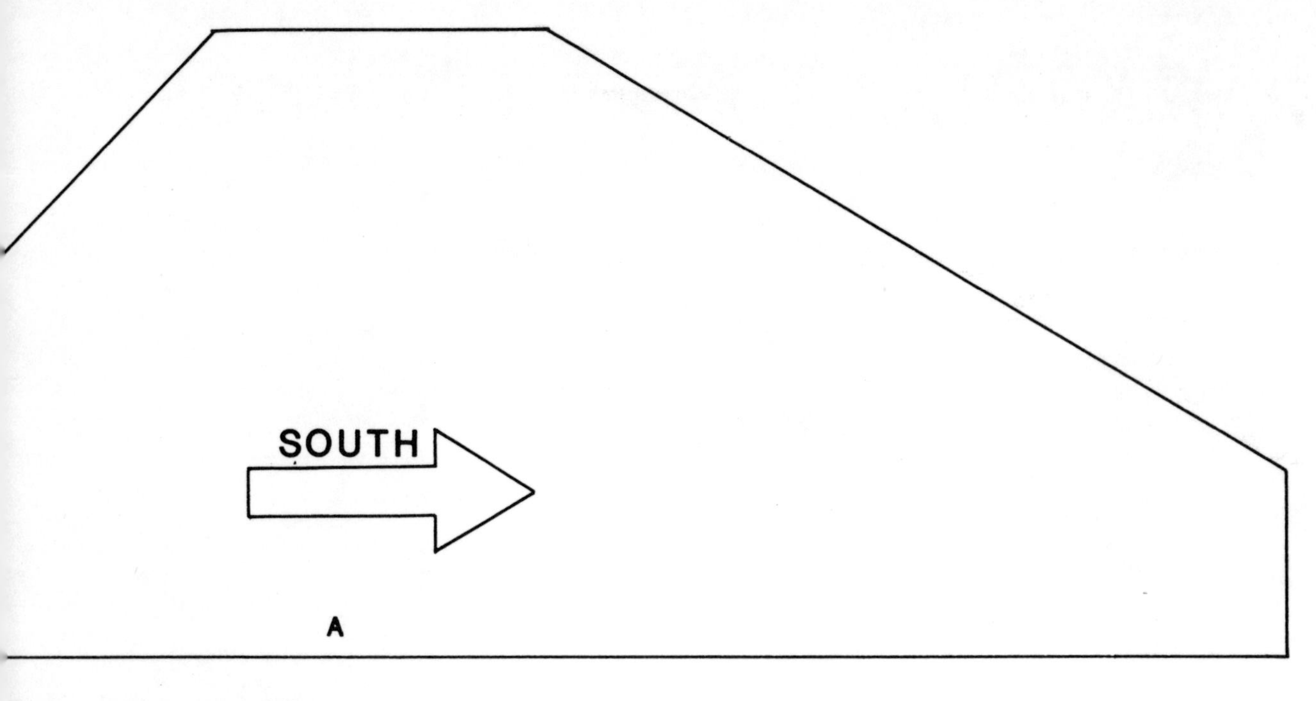

DE ASSEMBLY

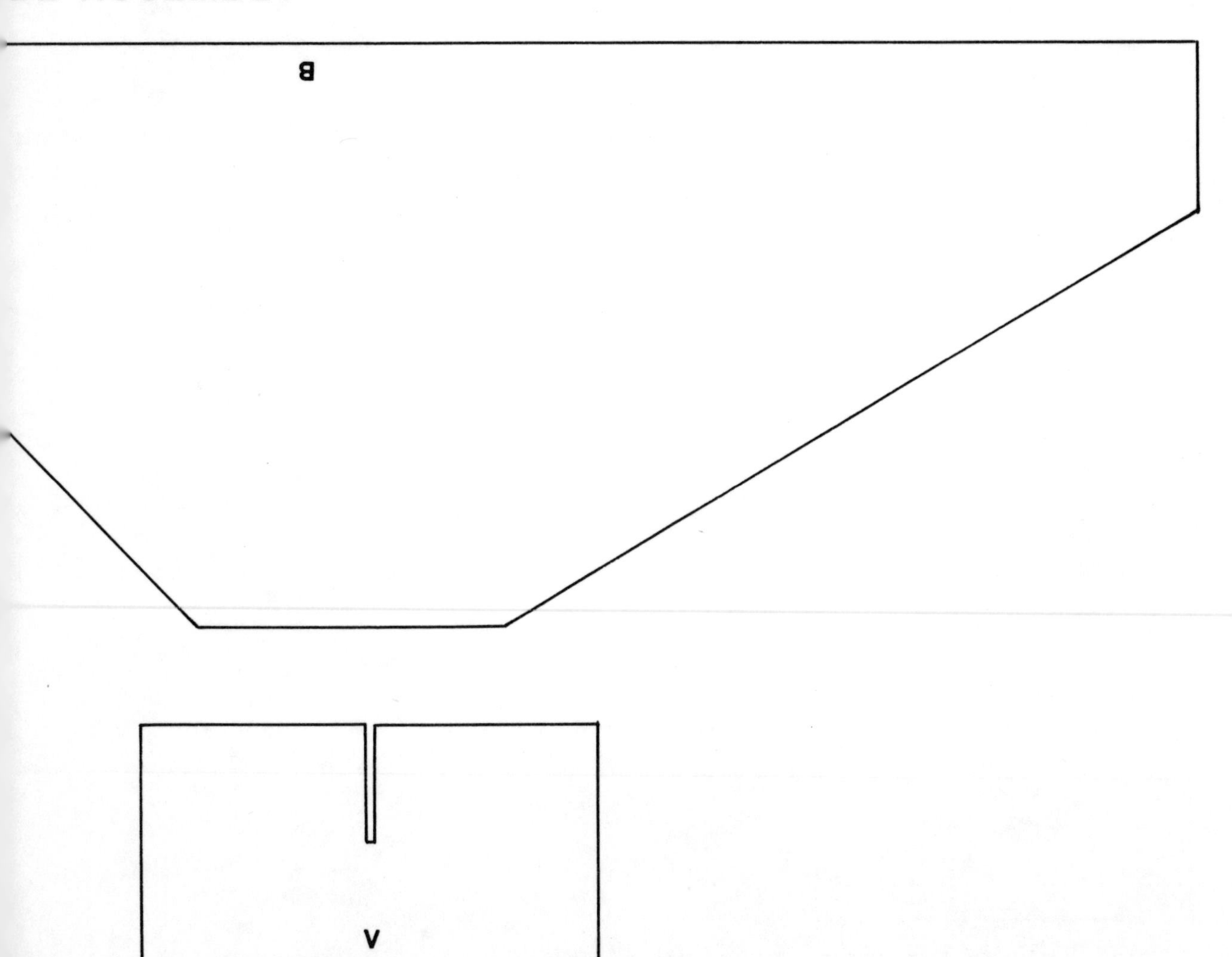

PIECE V

PARABOLIC SOLAR FURNACE

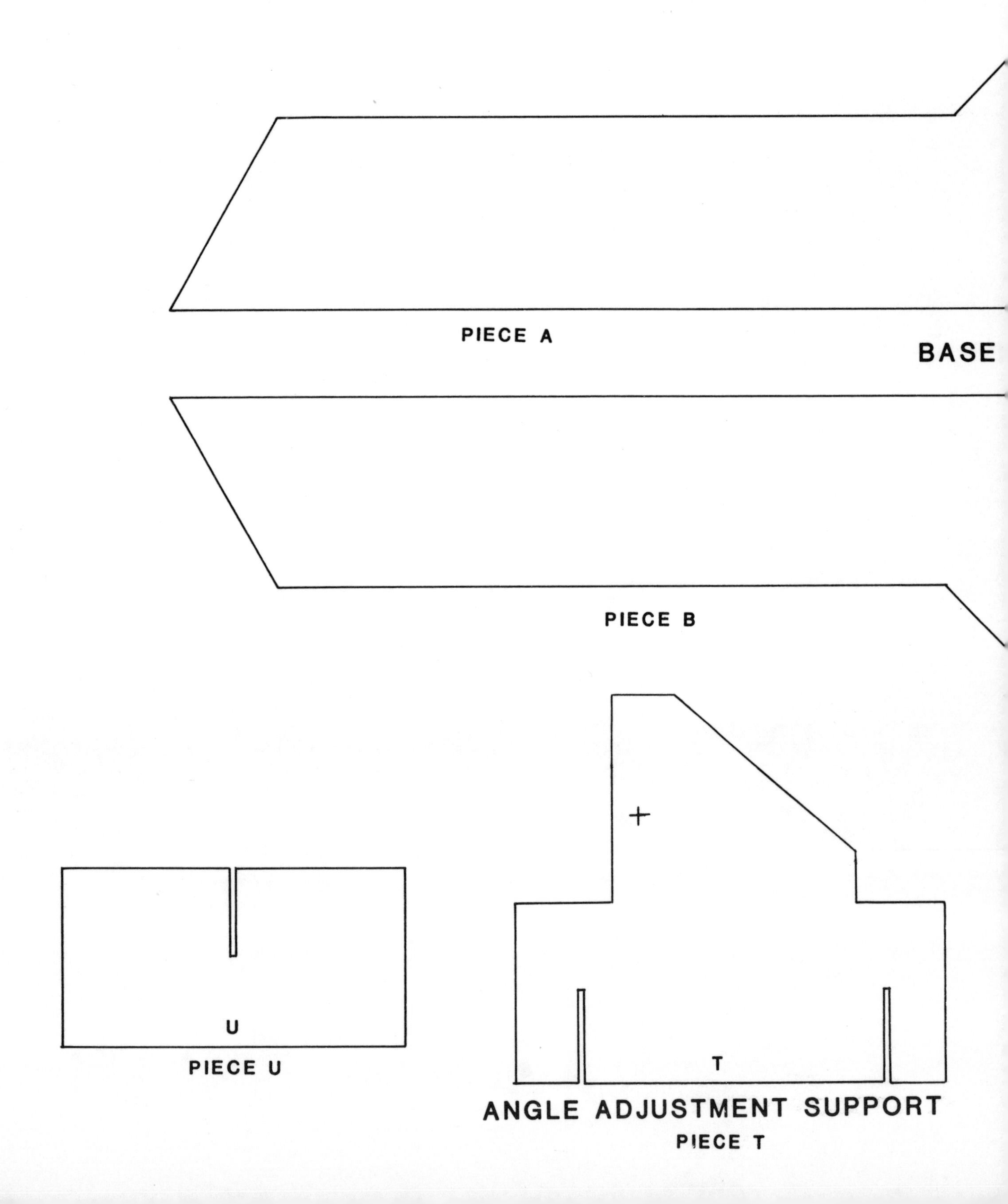

PROJECT
8

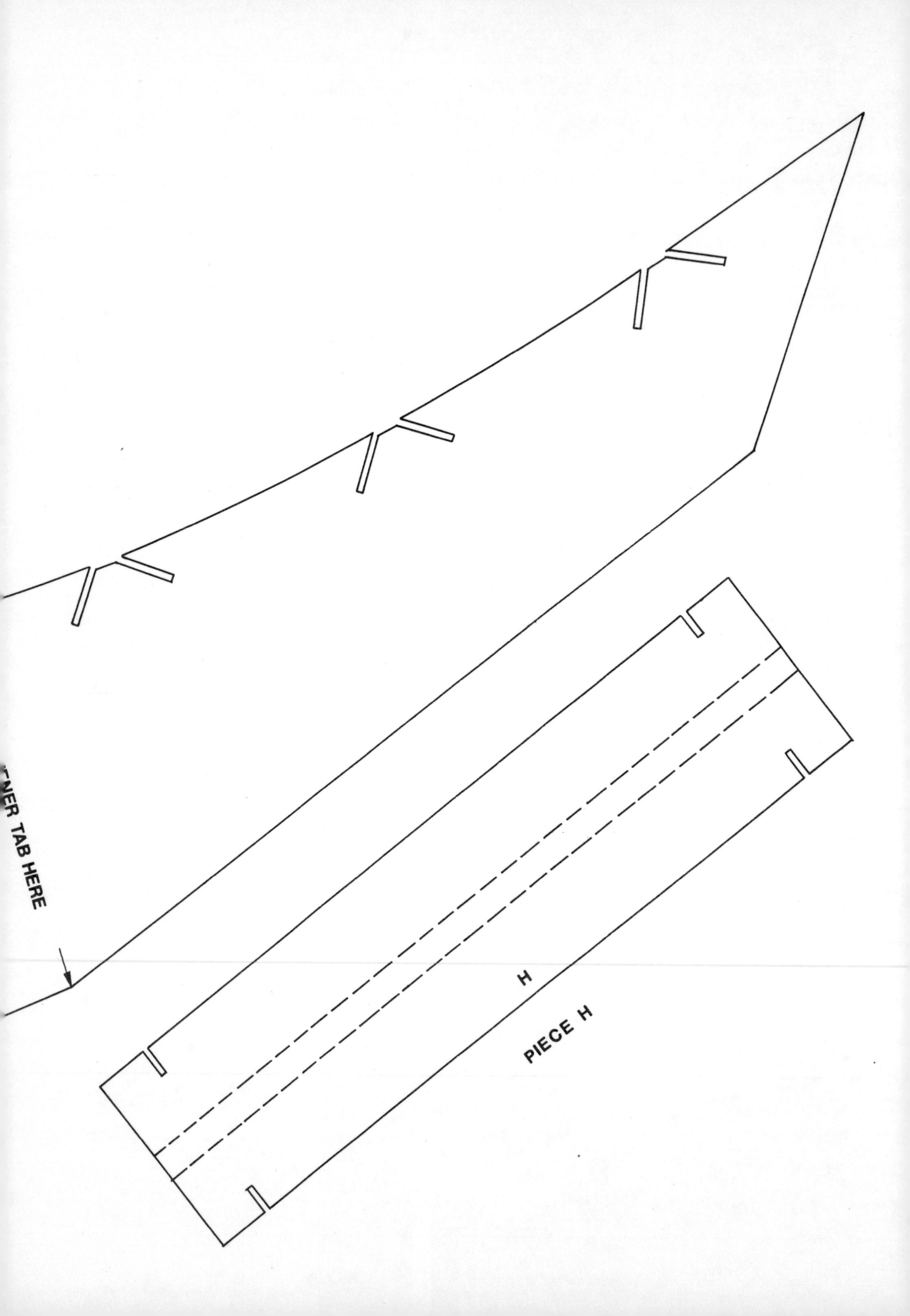
NER TAB HERE
H
PIECE H

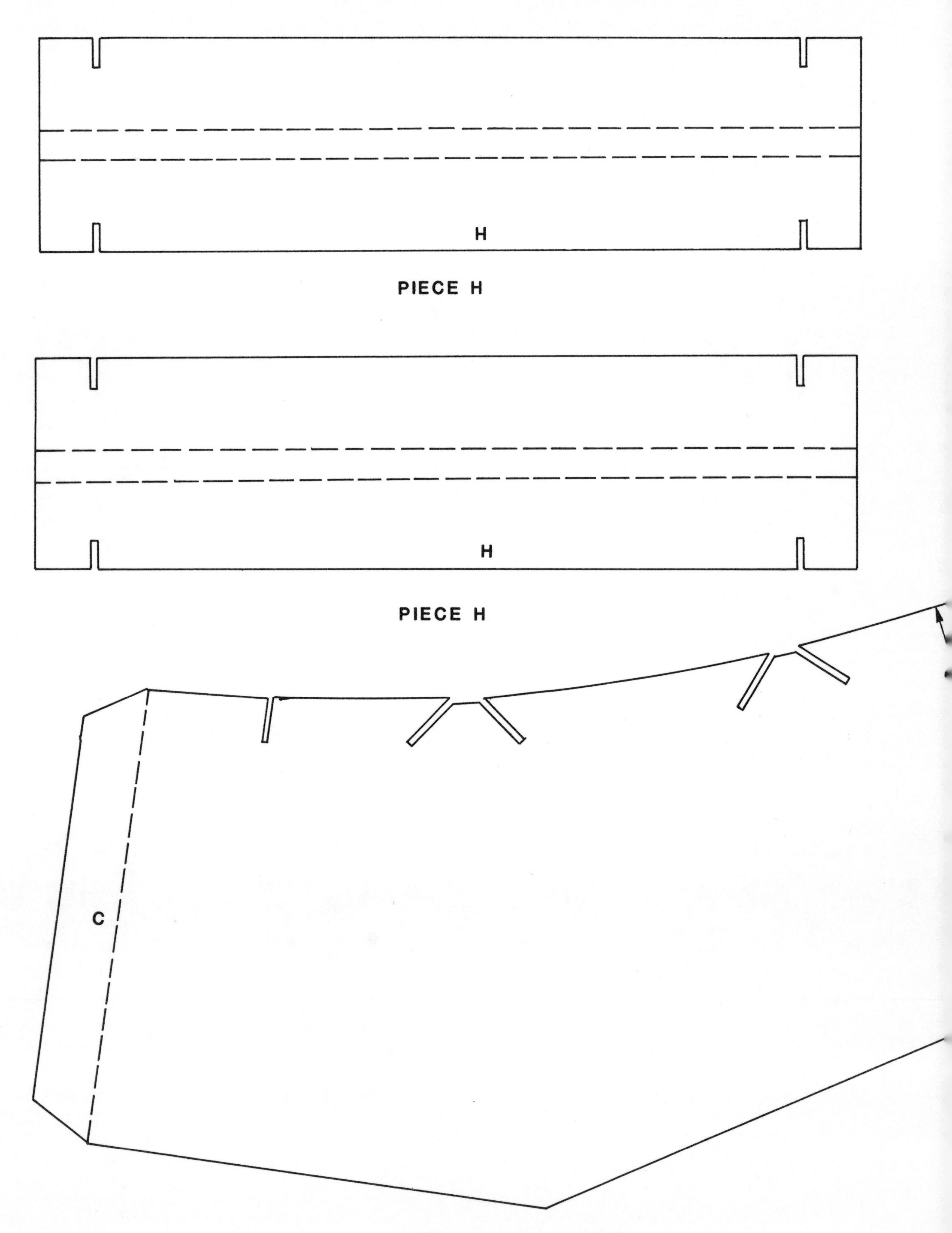

PARABOLA ARM

PIECE C

PROJECT
8

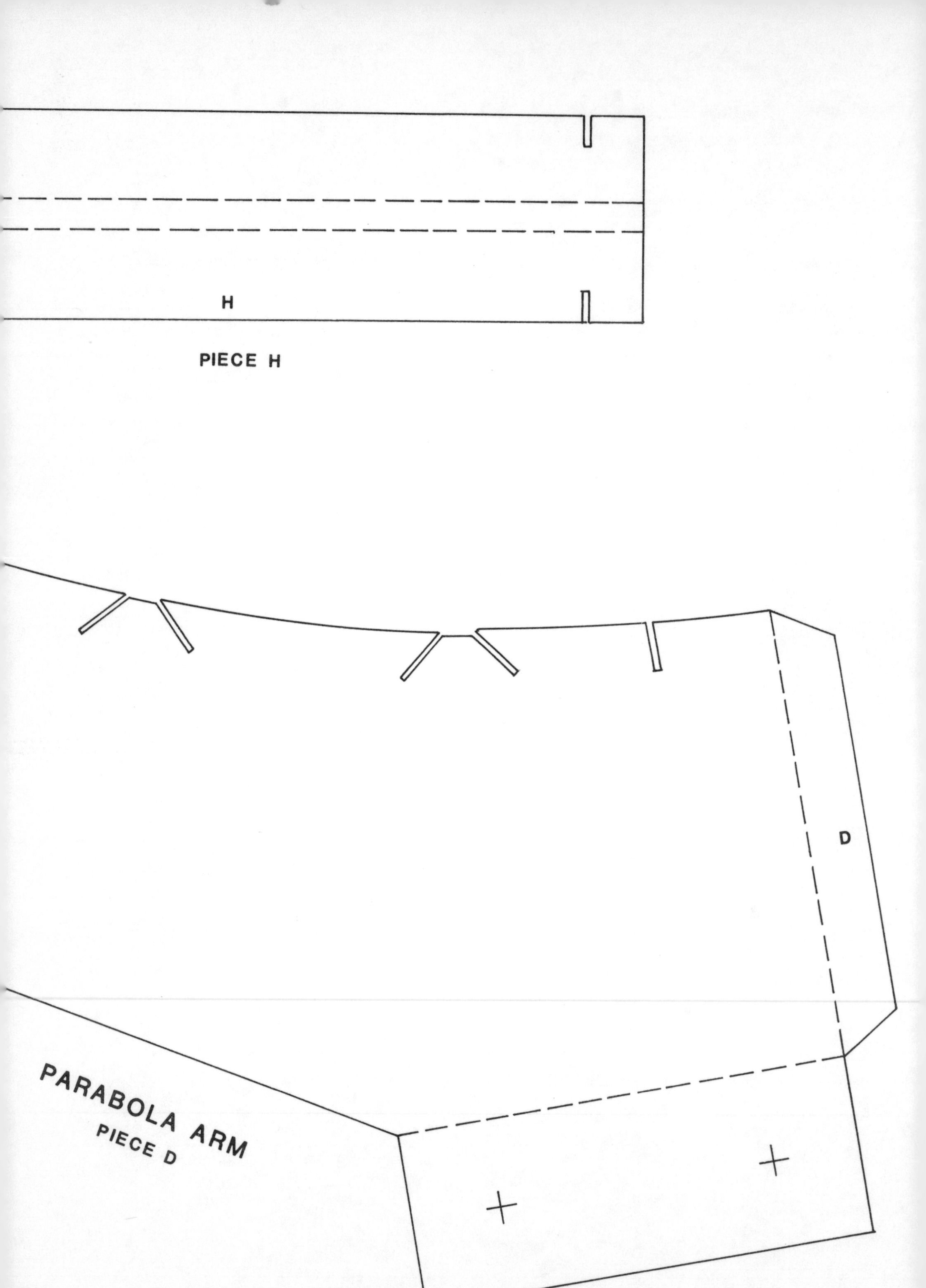
H
PIECE H
D
PARABOLA ARM
PIECE D

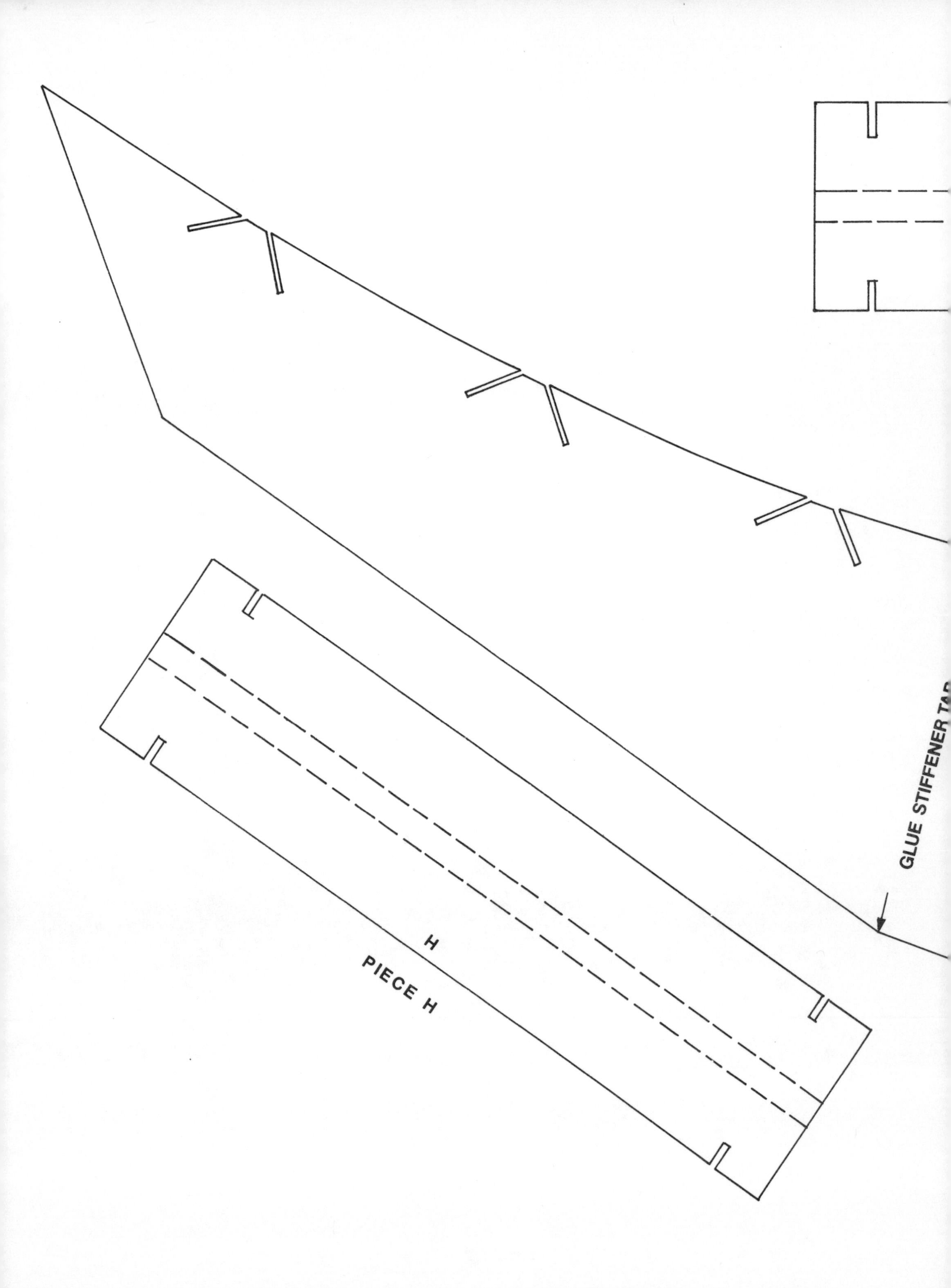
H
PIECE H
GLUE STIFFENER TAB

PROJECT
8

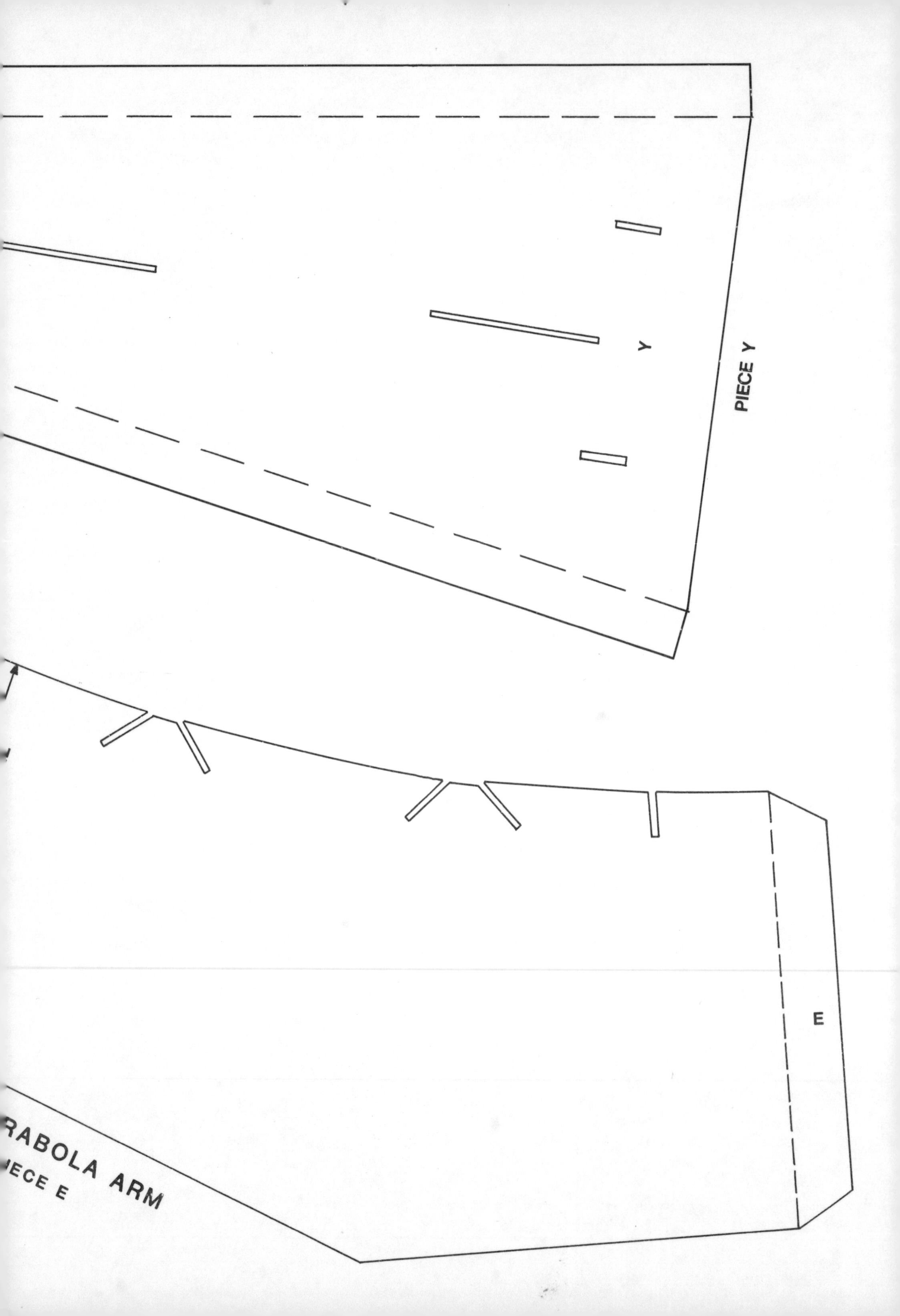
Y
PIECE Y
E
RABOLA ARM
IECE E

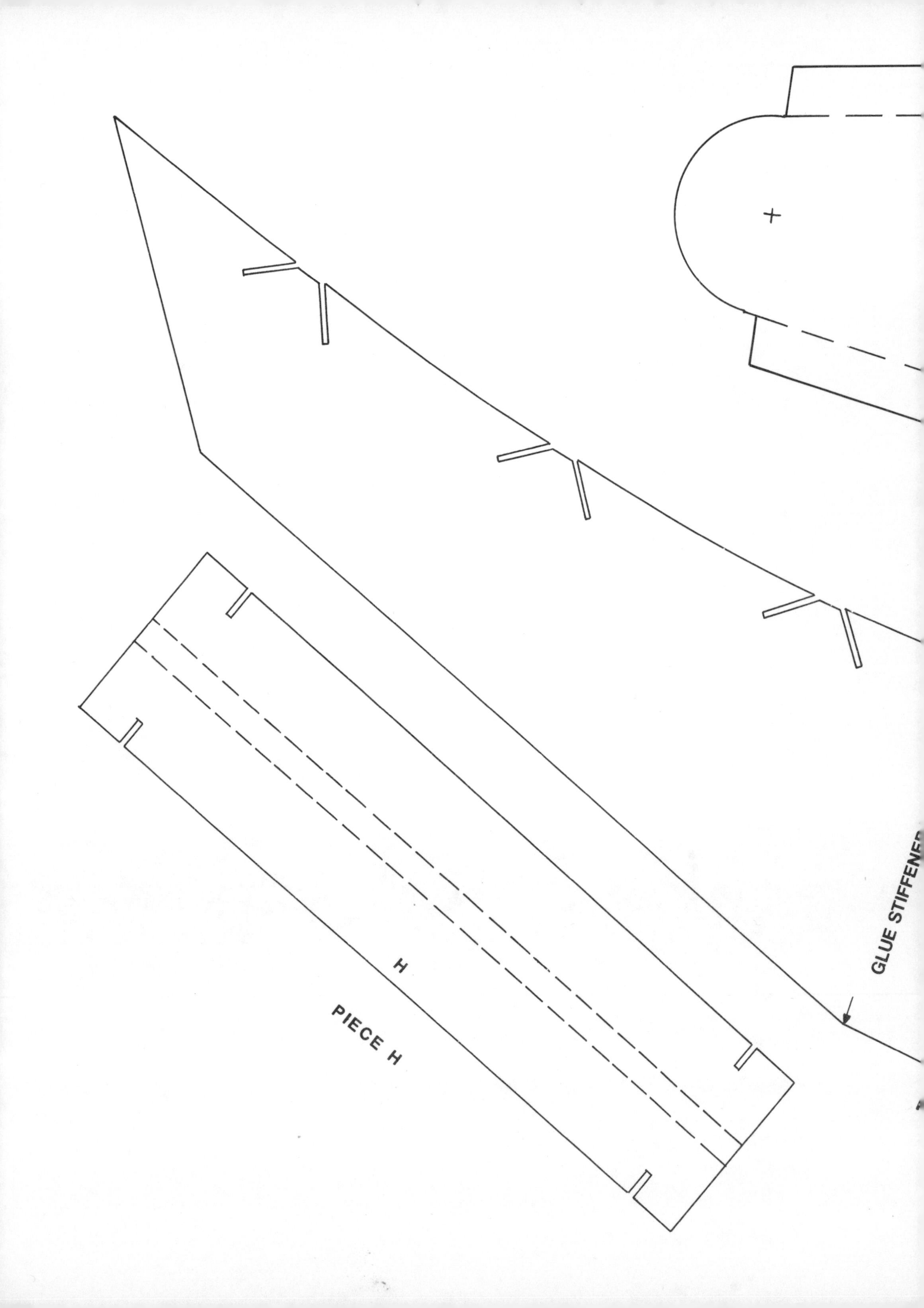

H
PIECE H
GLUE STIFFENER

PROJECT
8

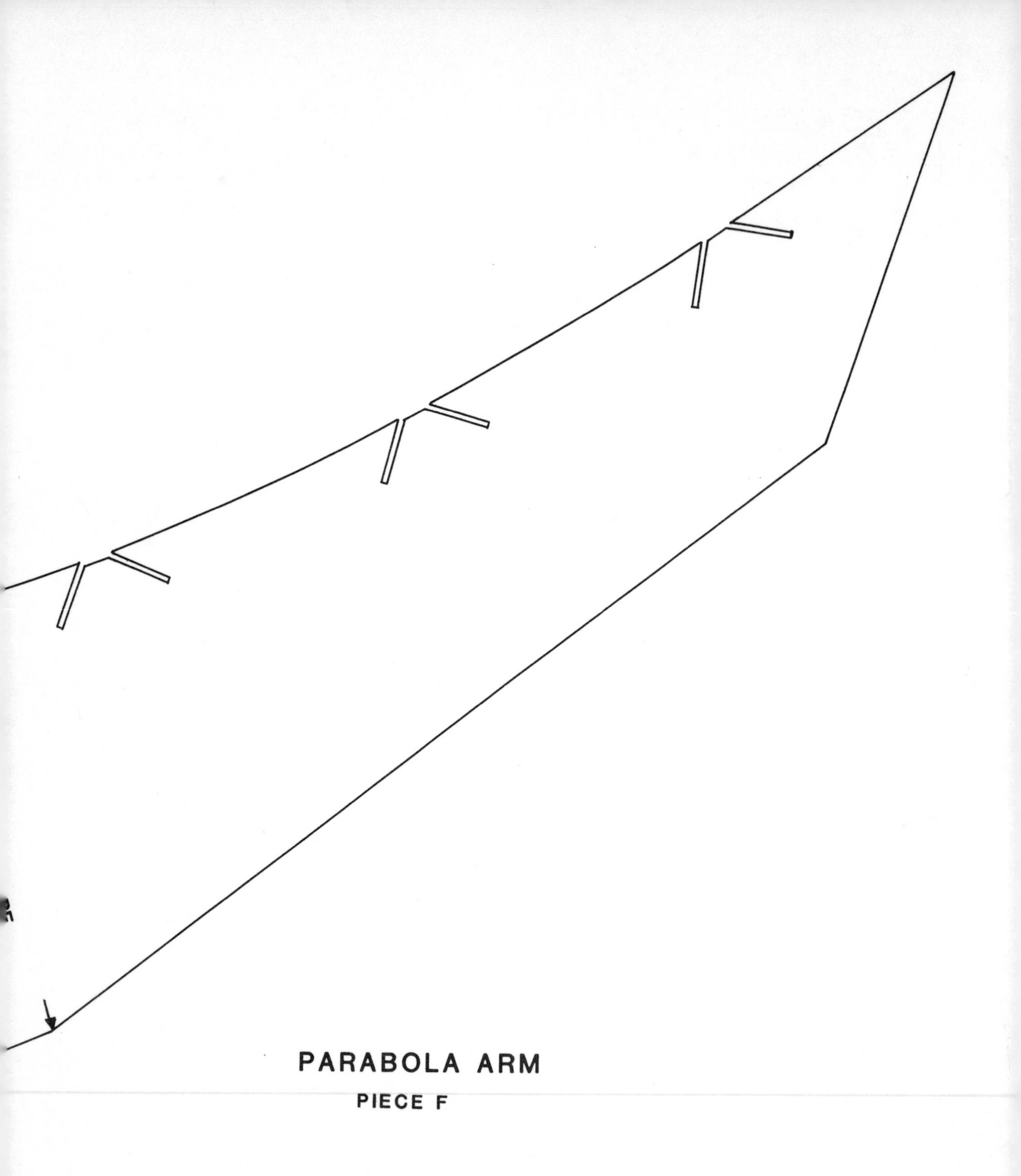

PARABOLA ARM

PIECE F

NOTE: FOCUS IS 7 1/16" HIGH

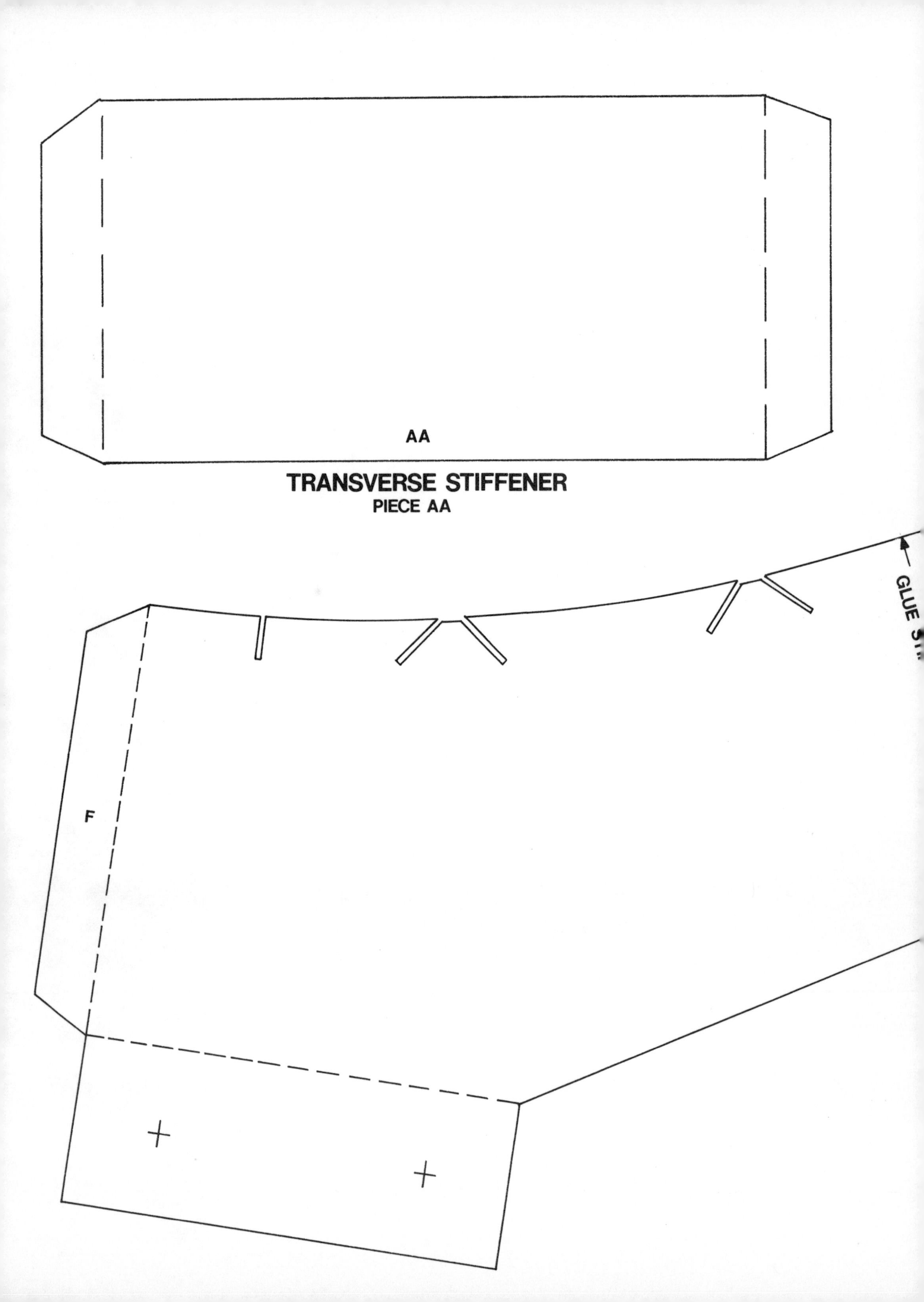
AA
TRANSVERSE STIFFENER
PIECE AA
GLUE ST
F

PROJECT
8

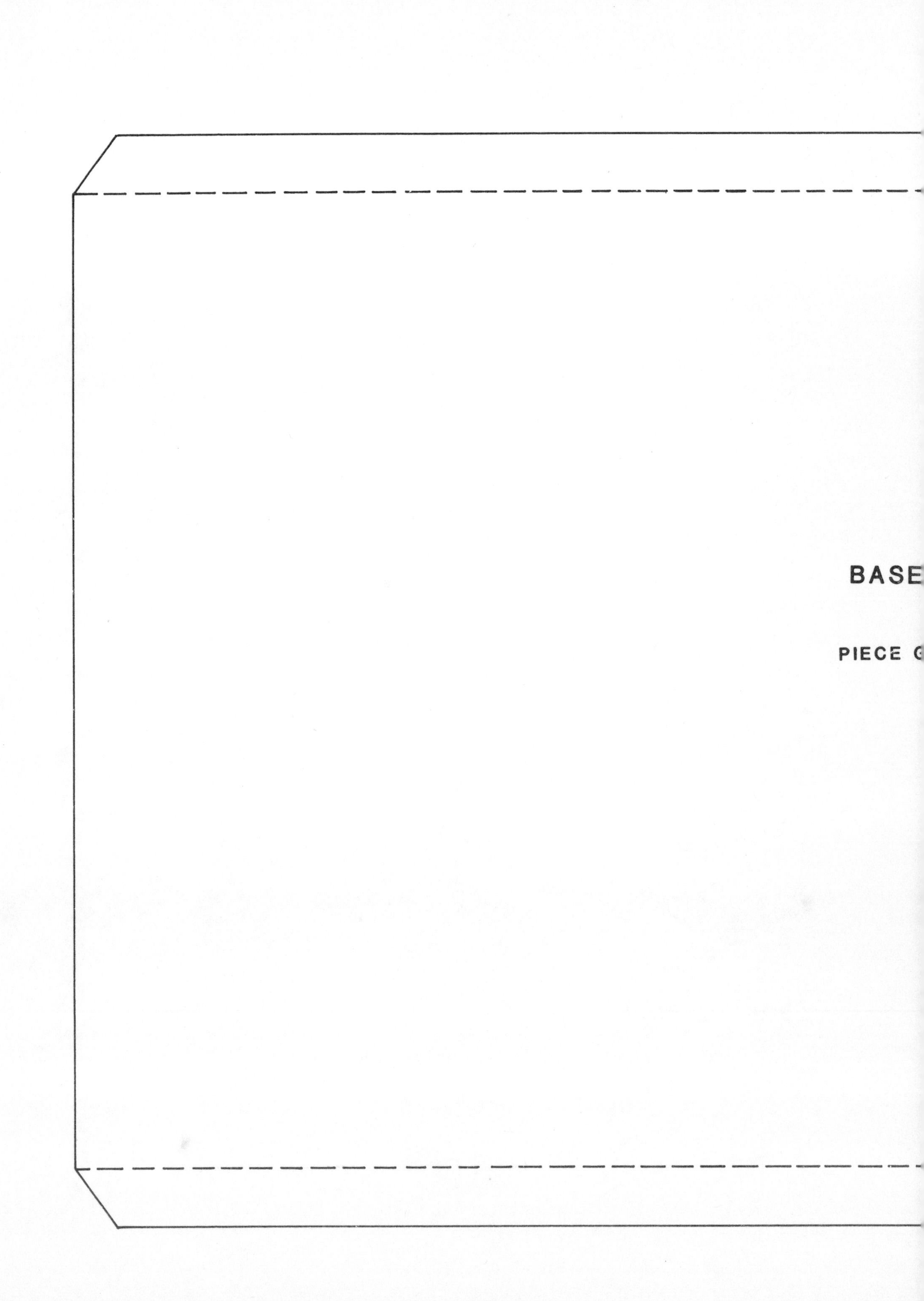
BASE
PIECE G

PROJECT
8

CURVE

ylar or aluminum foil)

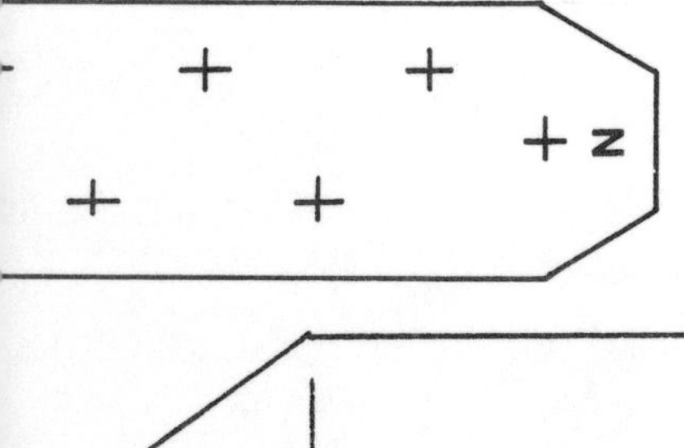

TRANSVERSE STIFFENER

PIECE BB

BB

PARABOL

(cover with reflective

PIECE

ANGLE ADJUSTMENT ARM

PIECE N

PROJECT
8

'RV.E

.ECTIVE MYLAR OR ALUMINUM FOIL)

TOP X

FOCUS ARM GUSSET LOCK

PIECE X

TOP W

FOCUS ARM GUSSET LOCK

PIECE W

PARABOLIC

(USING SPRAY ADHESIVE, COVER WITH R

PIECE M

PROJECT
8

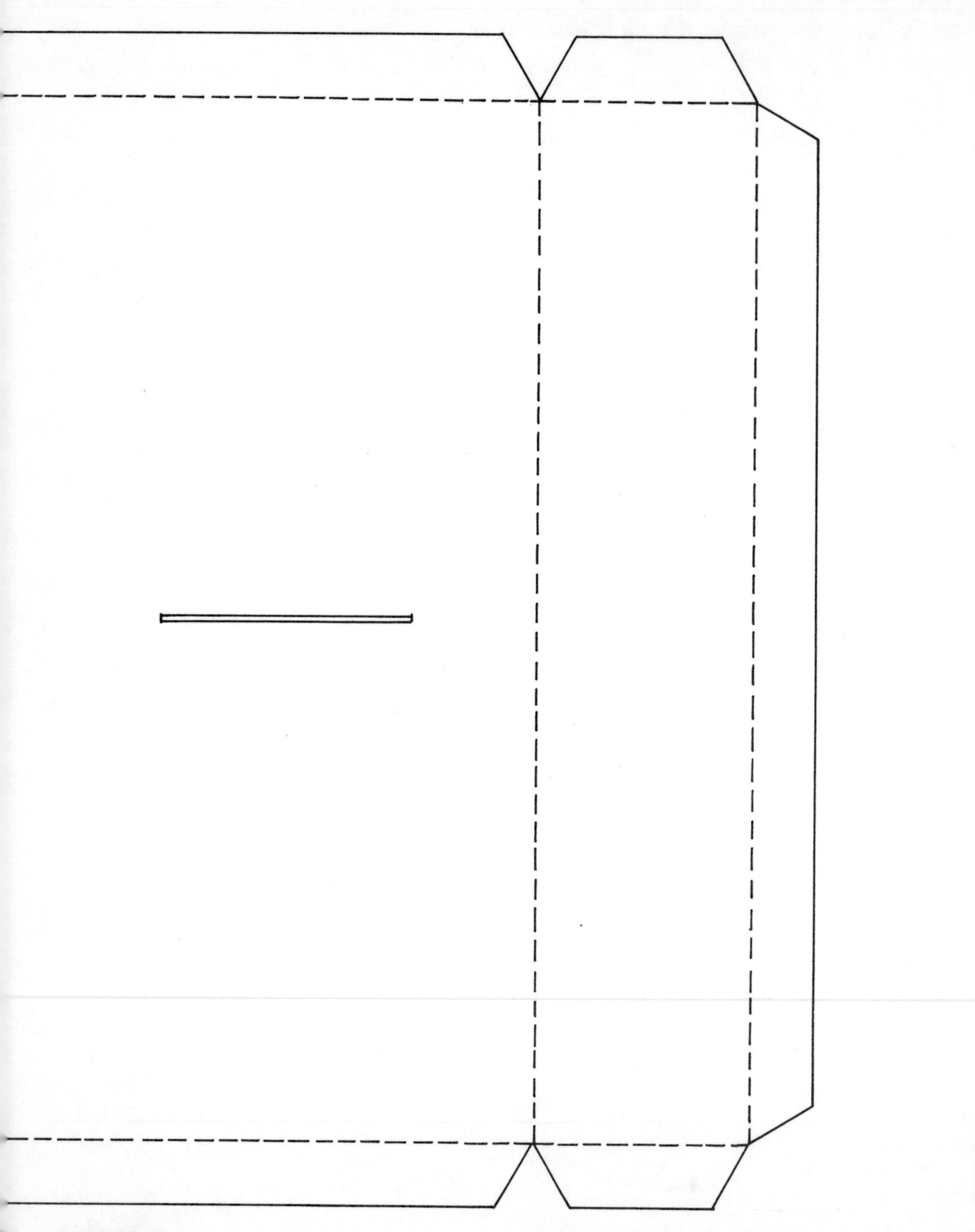

E ASSEMBLY

E Q

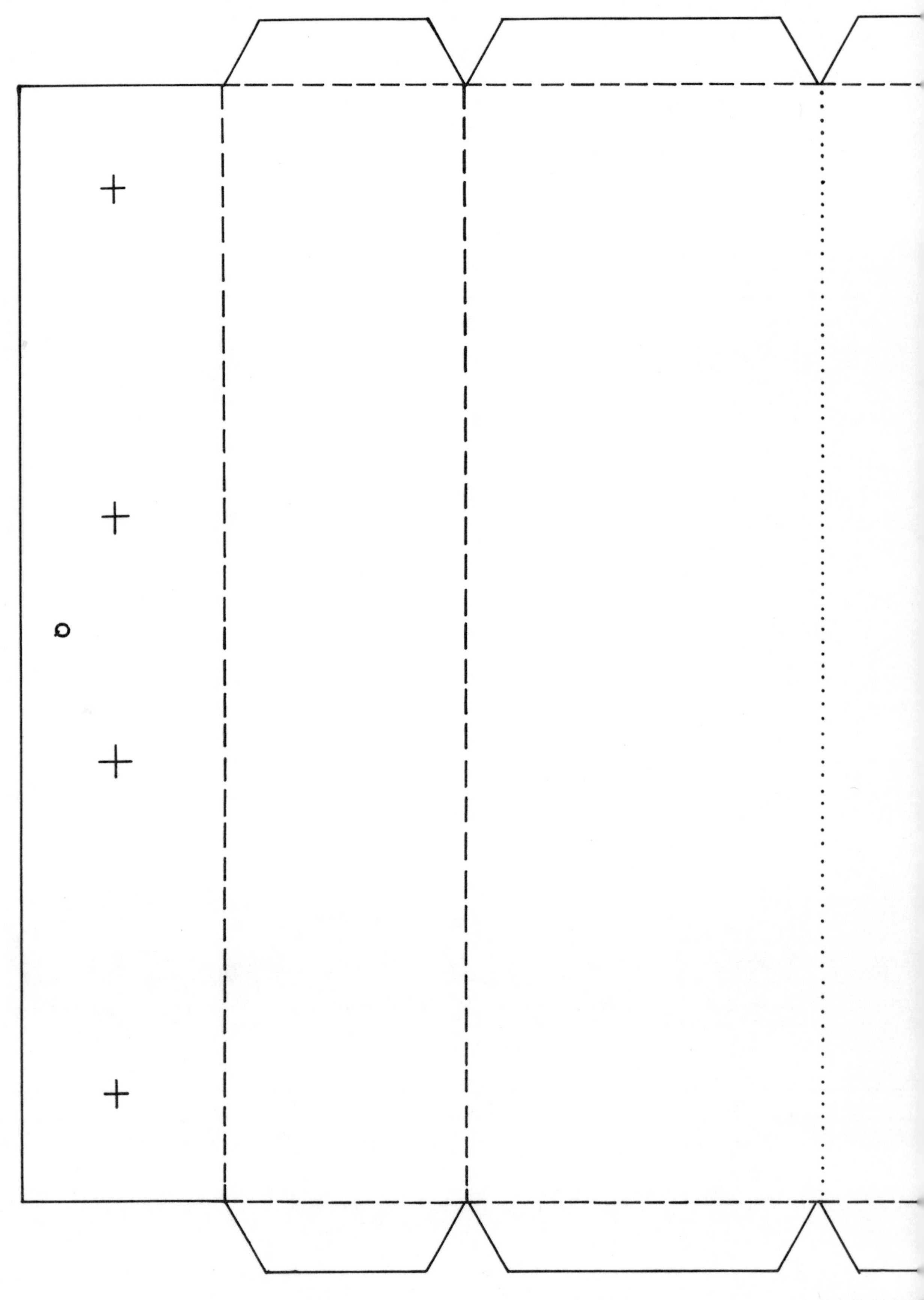

UPPER

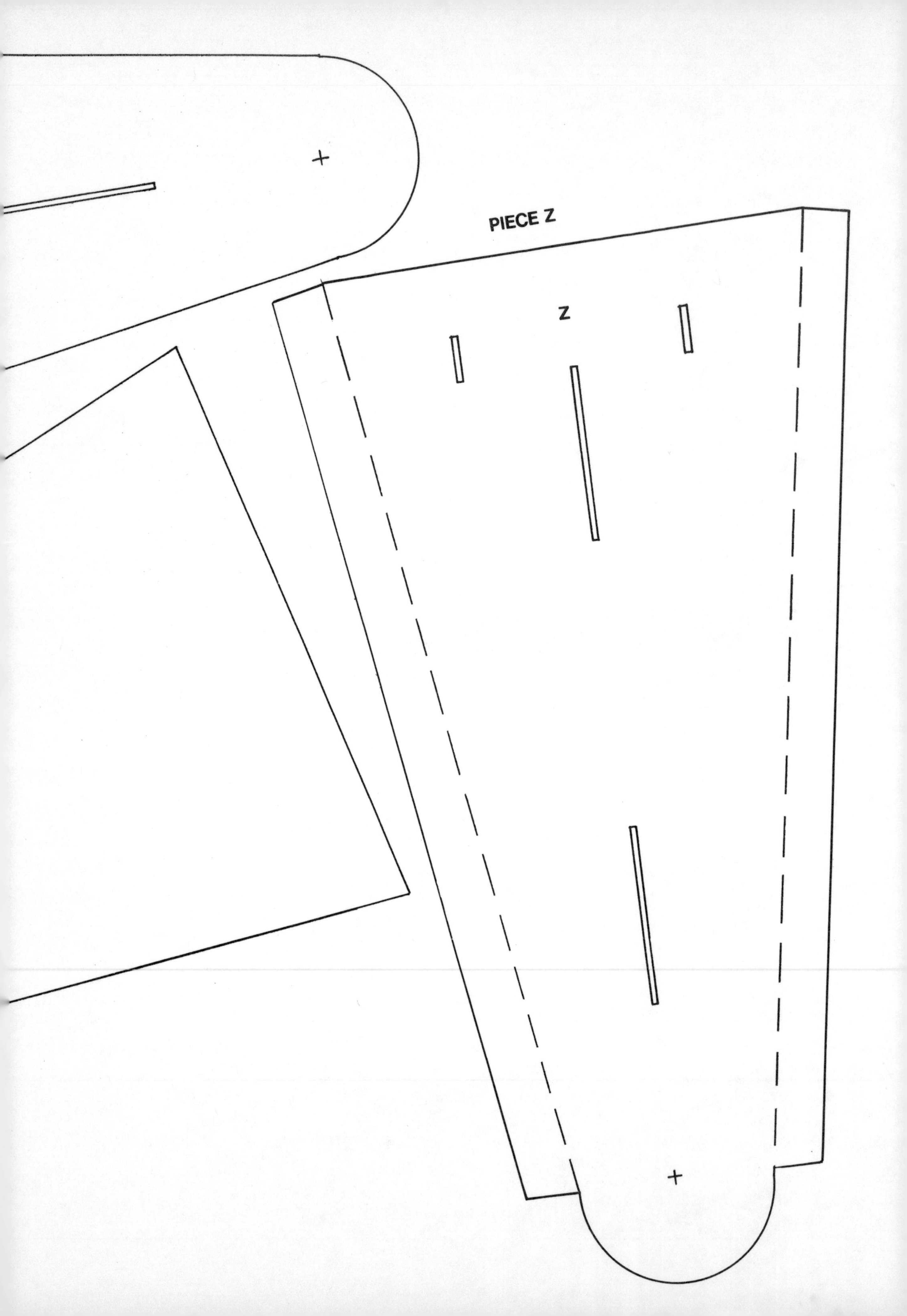
PIECE Z
Z

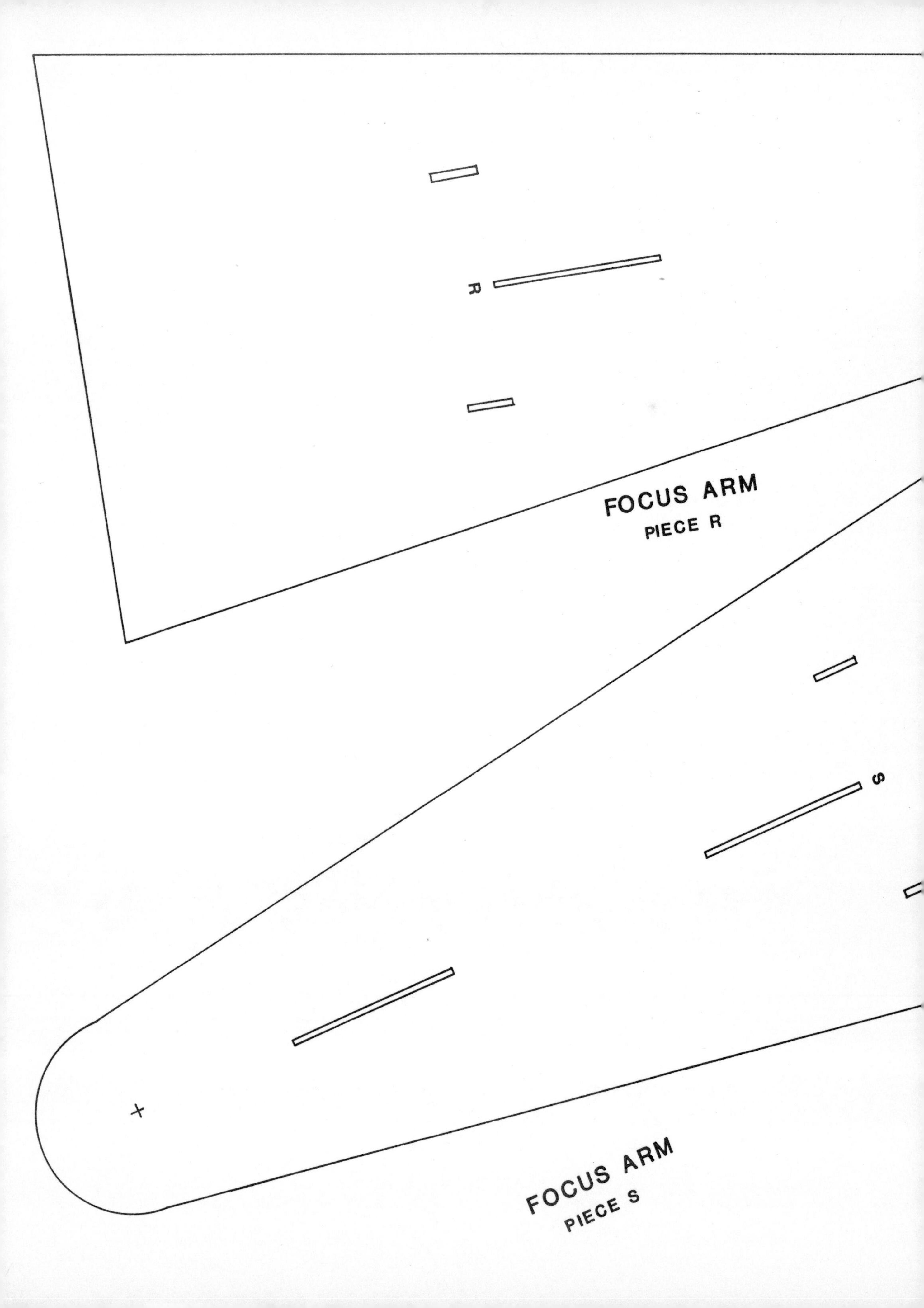
R
FOCUS ARM
PIECE R
S
FOCUS ARM
PIECE S

PROJECT 8

P

PARABOLIC SOLAR FURNACE

INSTRUCTIONS

1. PUT HOT DOG OR OTHER FOOD ITEM ON ROTISSERIE ROD.
2. ORIENT "SOUTH" ARROW TOWARD TRUE, OR CELESTIAL, SOUTH.
3. ADJUST THE ANGLE ADJUSTMENT ARM UNTIL SUNLIGHT FOCUSES ONTO FOOD.
4. SINCE THE SUN'S POSITION IS CONTINUALLY CHANGING, READJUST THE FOCUS ANGLE AS NEEDED.

NOTE: BE CAREFUL WHEN USING THIS SOLAR REFLECTOR: SUNLIGHT CAN PERMANENTLY DAMAGE THE EYES. WEAR PROTECTIVE DARK LENSES OR ELSE AVOID LOOKING INTO THE REFLECTIVE FOIL WHEN SUNLIGHT CAN REFLECT INTO THE EYES.

UPPER BASE ASSEMBLY

PIECE P

PIECE H

H

H

PIECE H

J

PARABOLIC ARM DIVIDER

PIECE J

PIECE H

H

H

PIECE H

K

PARABOLIC ARM DIVIDER
PIECE K

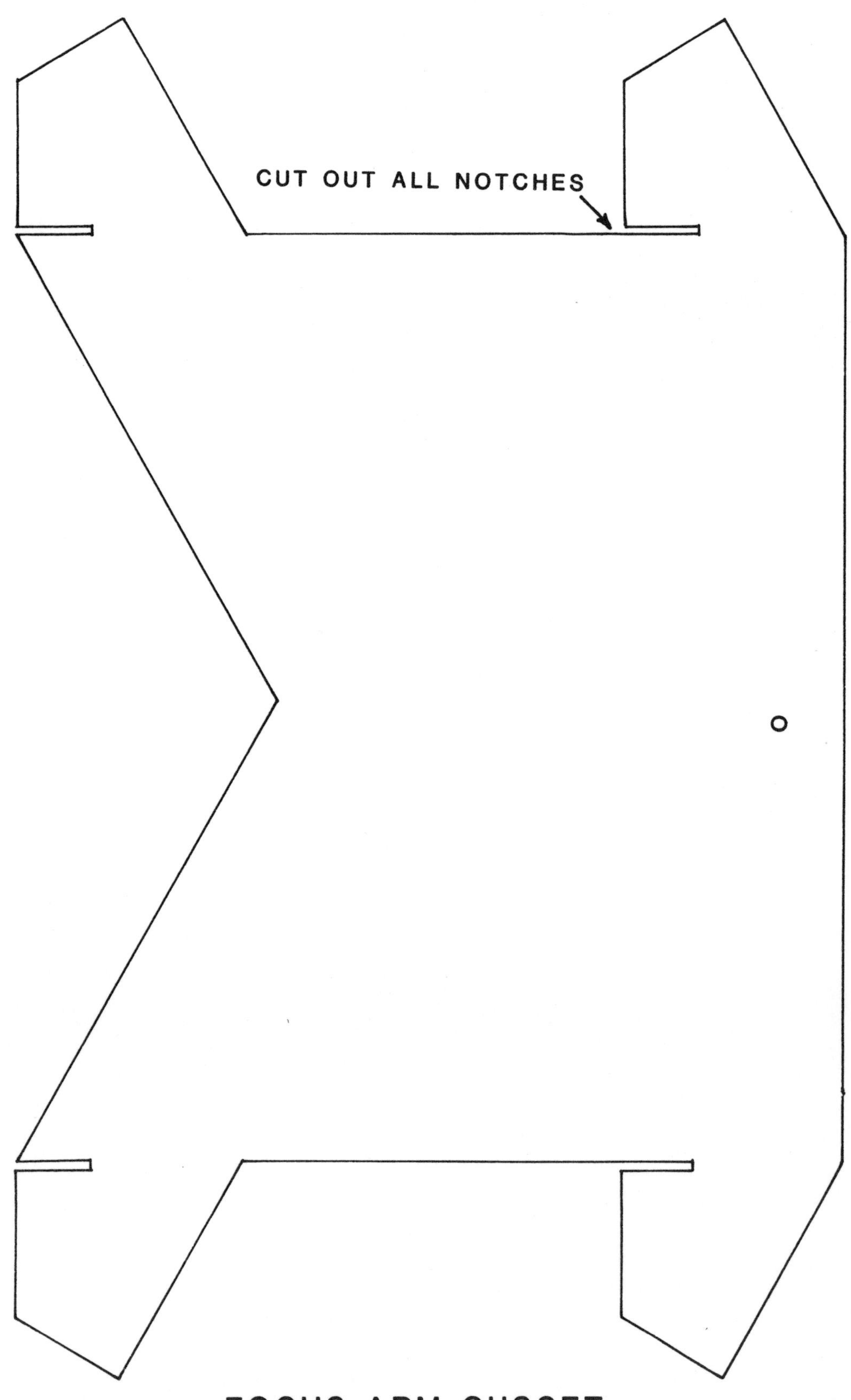

FOCUS ARM GUSSET

PIECE O

"SUNFLOWER"
PARABOLIC REFLECTOR

PROJECT 9

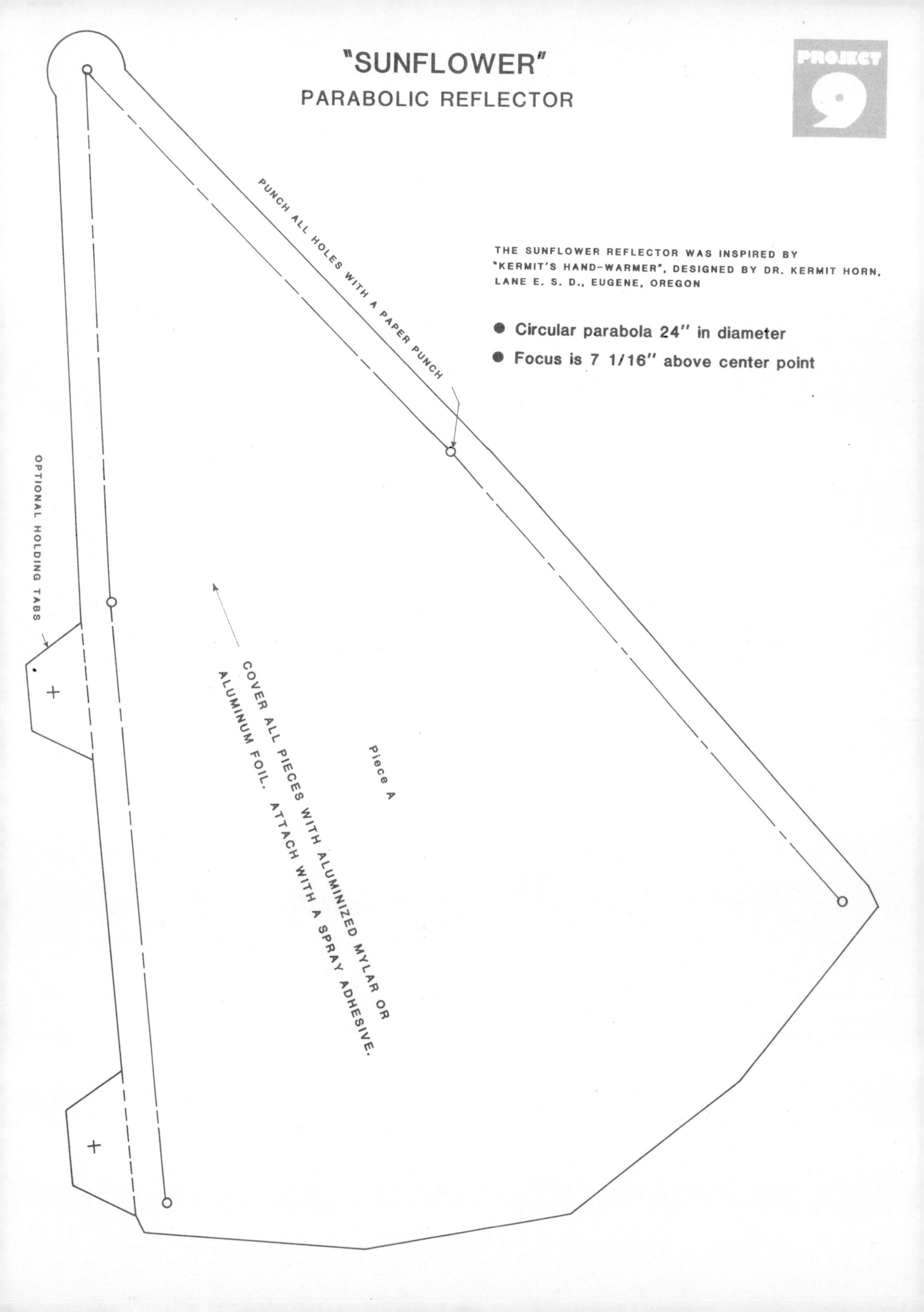

PUNCH ALL HOLES WITH A PAPER PUNCH
OPTIONAL HOLDING TABS
COVER ALL PIECES WITH ALUMINIZED MYLAR OR ALUMINUM FOIL. ATTACH WITH A SPRAY ADHESIVE.
Piece A

THE SUNFLOWER REFLECTOR WAS INSPIRED BY "KERMIT'S HAND-WARMER", DESIGNED BY DR. KERMIT HORN, LANE E. S. D., EUGENE, OREGON

- **Circular parabola 24″ in diameter**
- **Focus is 7 1/16″ above center point**

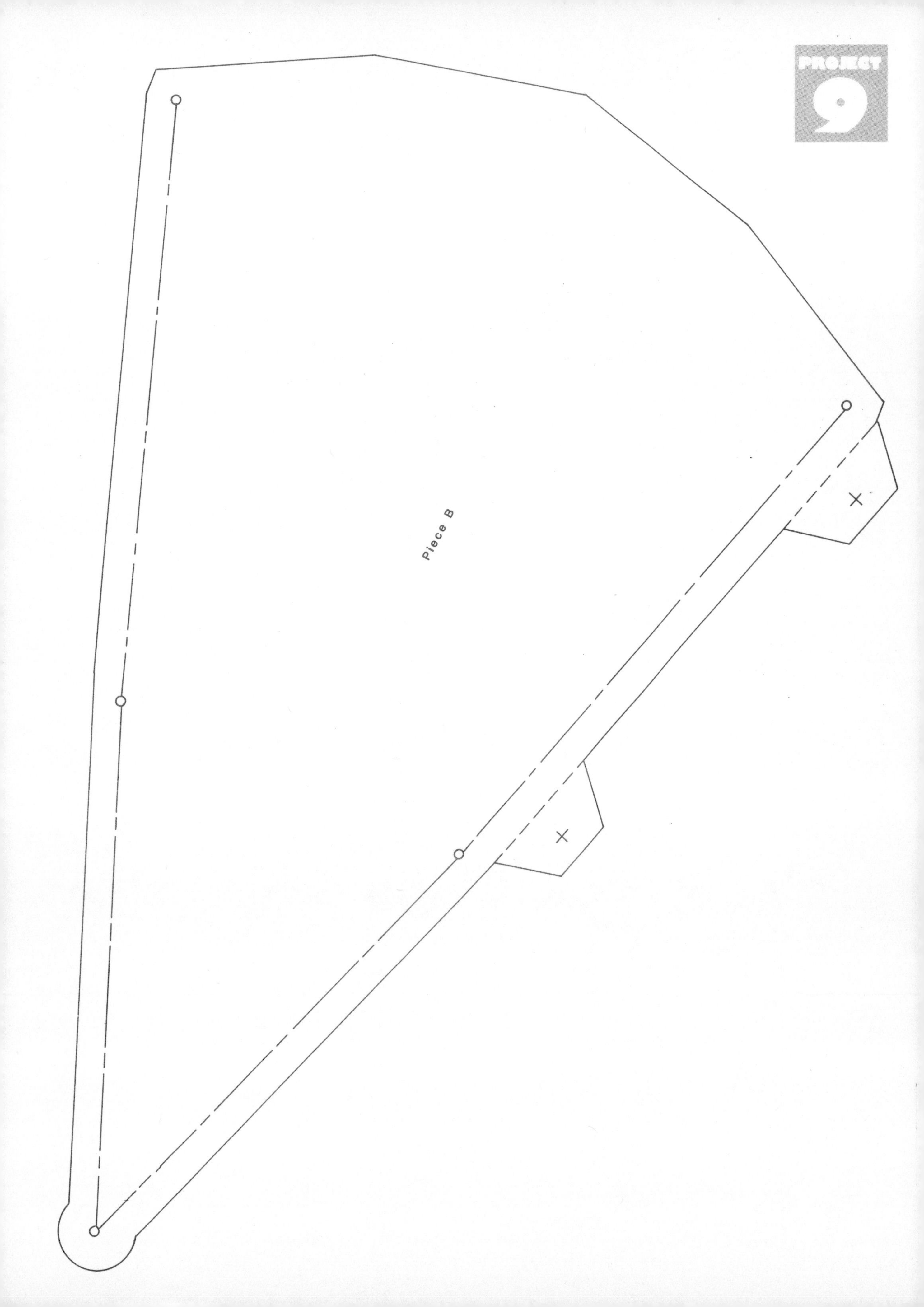
PROJECT
9
Piece B

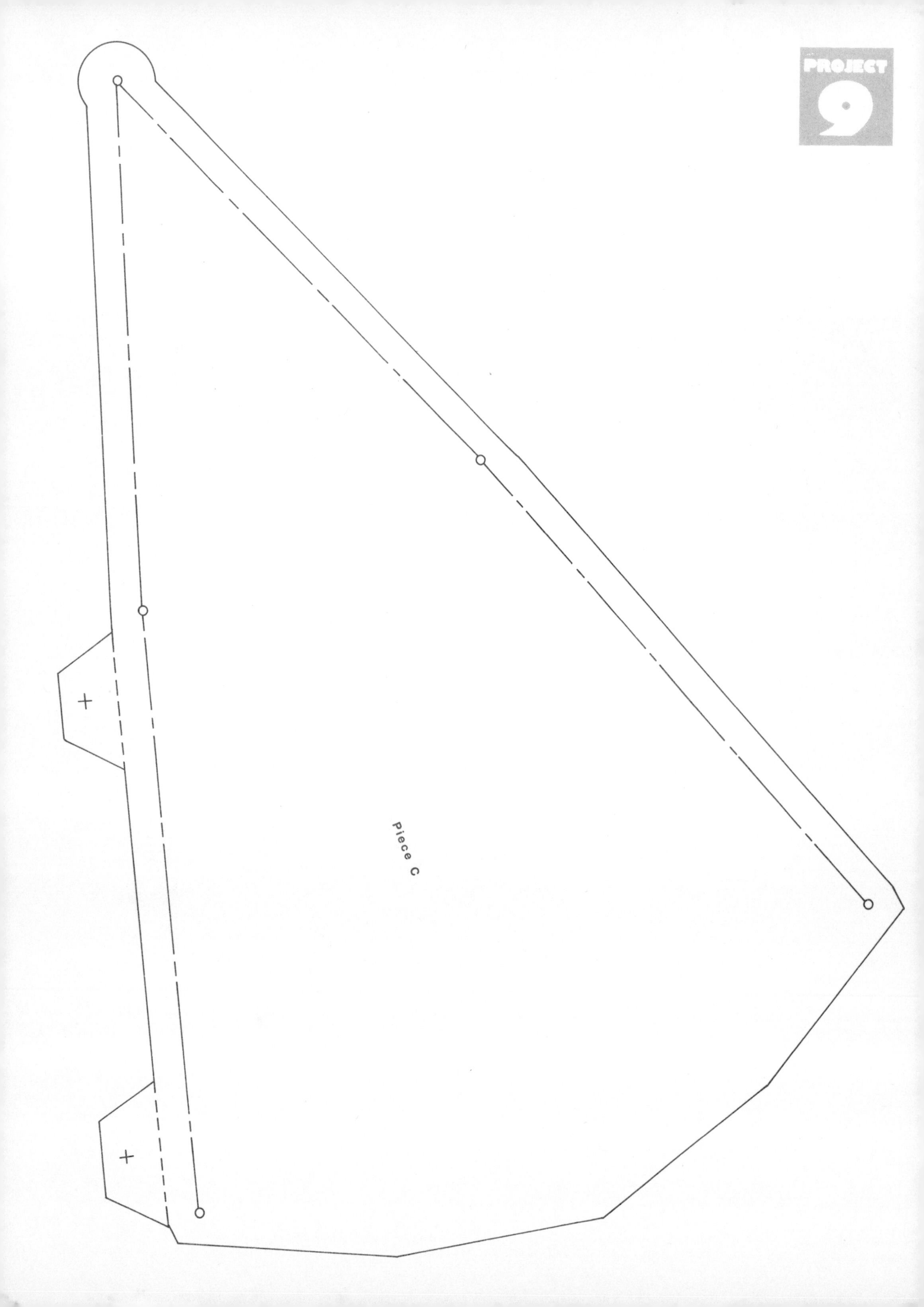
PROJECT
9
Piece C

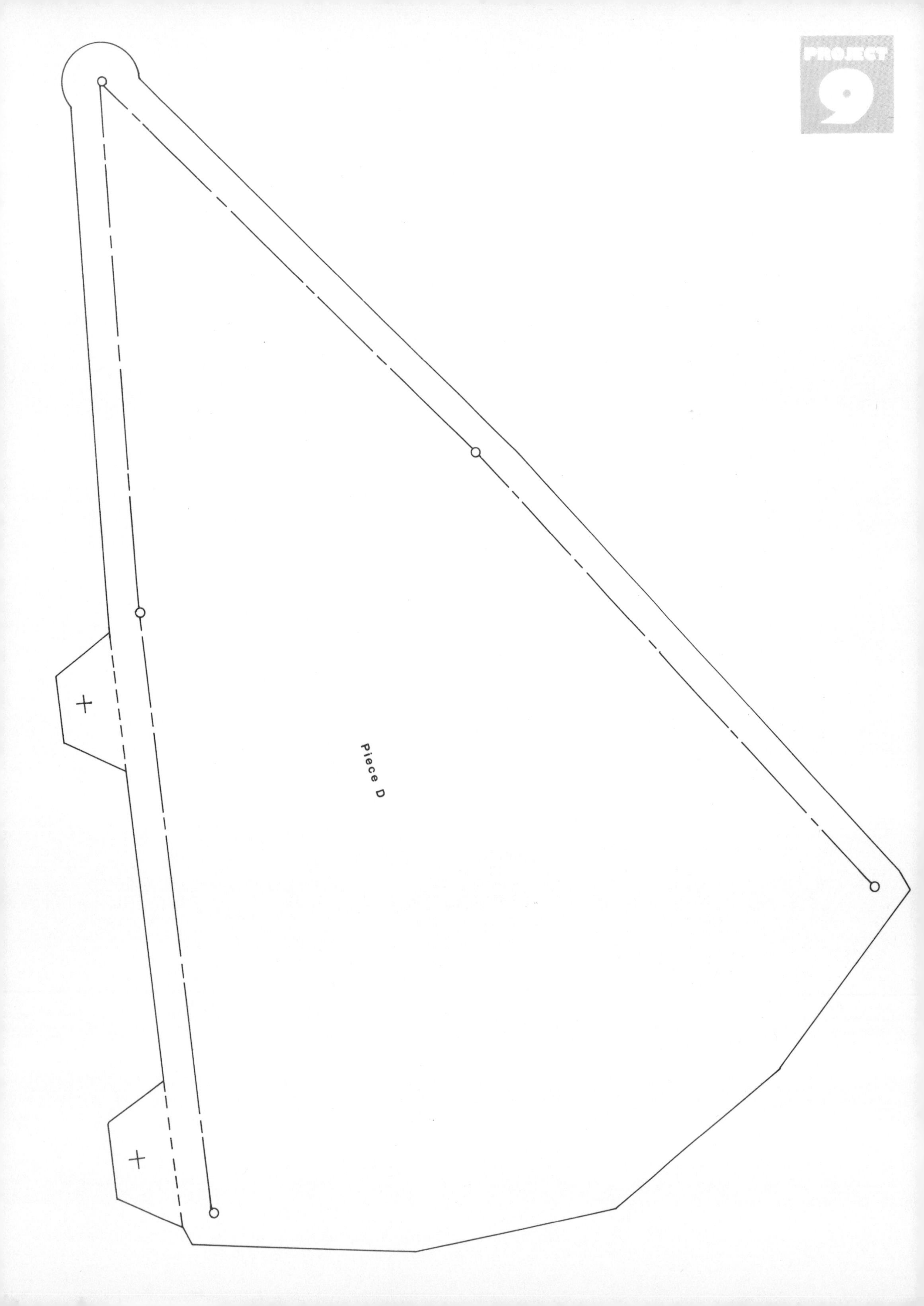
PROJECT
9
Piece D

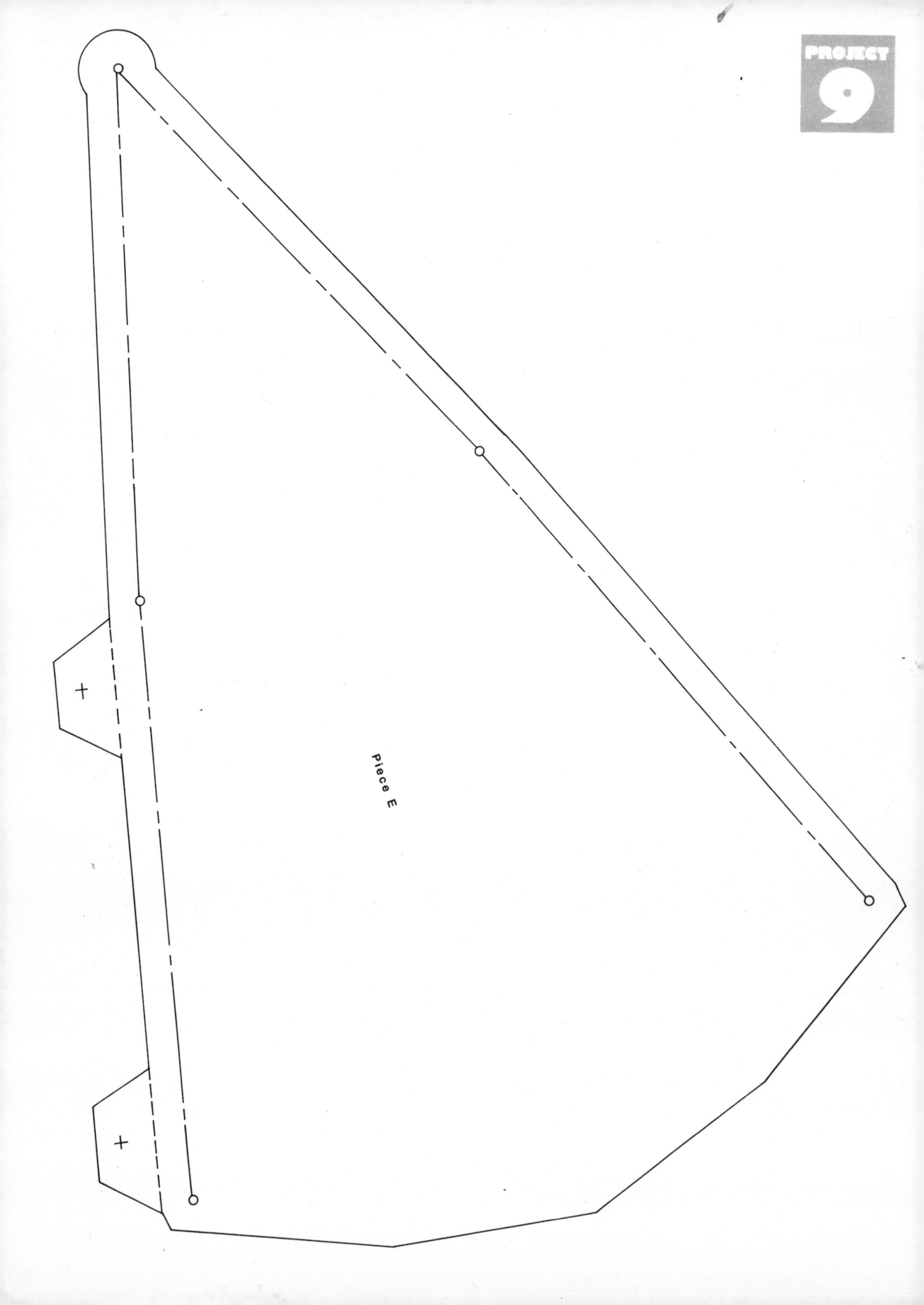
PROJECT 9
Piece E

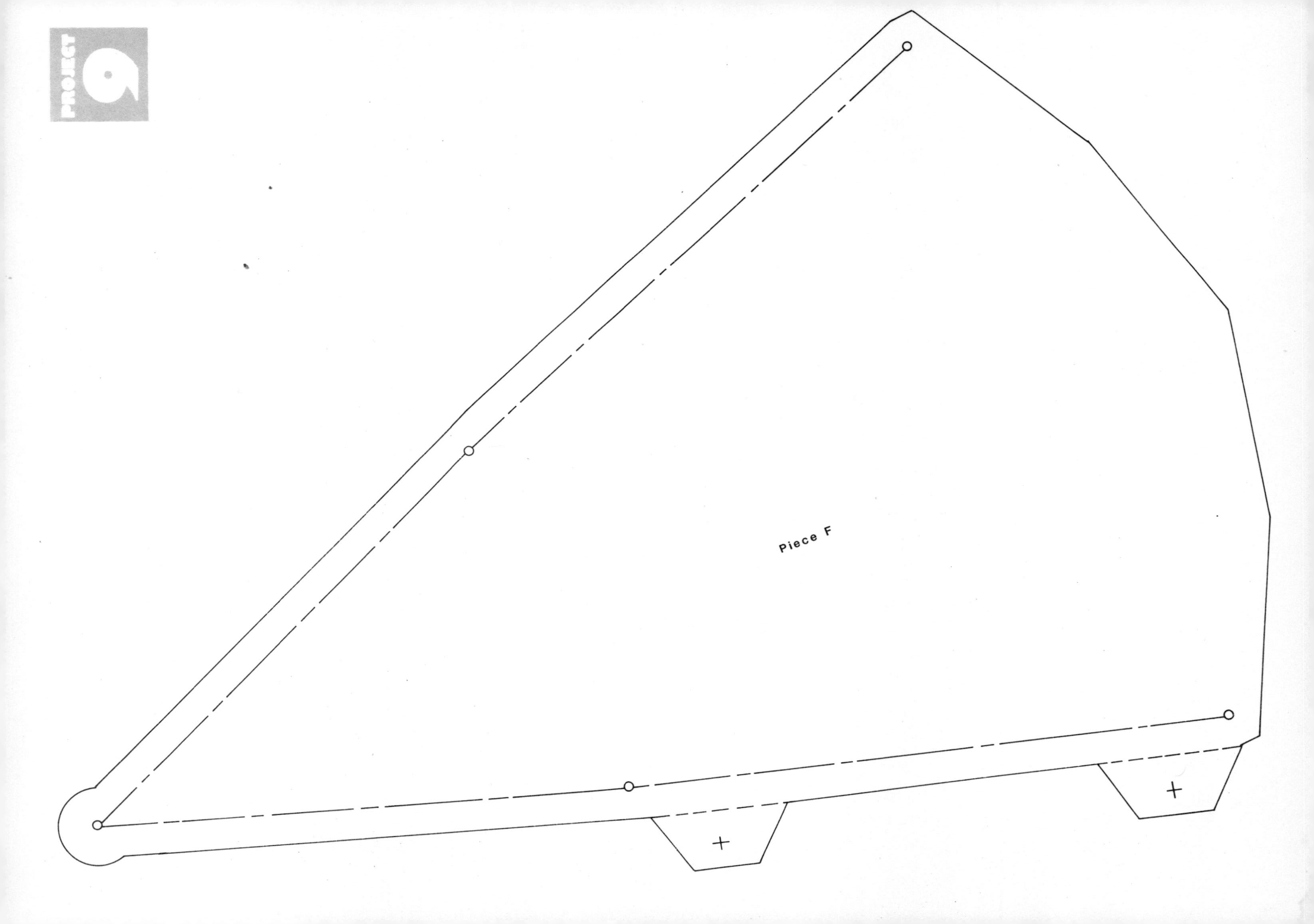
PROJECT
9
Piece F

Piece G

(THIS PIECE MUST BE ATTACHED NEXT-TO-LAST, SINCE IT HAS NO HOLDING TABS)